Interdisciplinary Applied Mathematics

Volume 32

Editors
S.S. Antman   **J.E. Marsden**   **L. Sirovich**

Geophysics and Planetary Sciences

Imaging, Vision, and Graphics

Mathematical Biology
L. Glass, J.D. Murray

Mechanics and Materials
R.V. Kohn

Systems and Control
S.S. Sastry, P.S. Krishnaprasad

Problems in engineering, computational science, and the physical and biological sciences are using increasingly sophisticated mathematical techniques. Thus, the bridge between the mathematical sciences and other disciplines is heavily traveled. The correspondingly increased dialog between the disciplines has led to the establishment of the series: *Interdisciplinary Applied Mathematics*.

The purpose of this series is to meet the current and future needs for the interaction between various science and technology areas on the one hand and mathematics on the other. This is done, firstly, by encouraging the ways that mathematics may be applied in traditional areas, as well as point toward new and innovative areas of applications; and, secondly, by encouraging other scientific disciplines to engage in a dialog with mathematicians outlining their problems to both access new methods and suggest innovative developments within mathematics itself.

The series will consist of monographs and high-level texts from researchers working on the interplay between mathematics and other fields of science and technology.

Interdisciplinary Applied Mathematics

Volumes published are listed at the end of the book.

Dominik Wodarz

Killer Cell Dynamics
Mathematical and Computational Approaches to Immunology

Dominik Wodarz
Department of Ecology and Evolution
University of California
321 Steinhaus Hall, Mail Code 2525
Irvine, CA 92697-2525
USA
dwodarz@uci.edu

Editors
S.S. Antman
Department of Mathematics and
Institute for Physical Science and
  Technology
University of Maryland
College Park, MD 20742
USA
ssa@math.umd.edu

J.E. Marsden
Control and Dynamical Systems
Mail Code 107-81
California Institute of Technology
Padadena, CA 91125
USA
marsden@cds.caltech.edu

L. Sirovich
Division of Applied Mathematics
Brown University
Providence, RI 02912
USA
chico@camelot.mssm.edu

Cover illustration: *The Killing*, painting by Natalia Komarova

Mathematics Subject Classification (2000): 37N25, 46N60, 62P10, 92-xx, 92Bxx, 92B05

Library of Congress Control Number: 2006931001

ISBN-10: 0-387-30893-8
ISBN-13: 978-0-387-30893-7

Printed on acid-free paper.

© 2007 Springer Science+Business Media, LLC
All rights reserved. This work may not be translated or copied in whole or in part without the written permission of the publisher (Springer Science+Business Media, LLC, 233 Spring Street, New York, NY 10013, USA), except for brief excerpts in connection with reviews or scholarly analysis. Use in connection with any form of information storage and retrieval, electronic adaptation, computer software, or by similar or dissimilar methodology now known or hereafter developed is forbidden.
The use in this publication of trade names, trademarks, service marks, and similar terms, even if they are not identified as such, is not to be taken as an expression of opinion as to whether or not they are subject to proprietary rights.

9 8 7 6 5 4 3 2 1

springer.com

*To
my parents Klara and Hans-Walter, my brother Andi, Pinktche,
my wife Natasha, and our children Sophia and Paulina. Paulina
was born during the final stages of writing this book.*

Preface

Systems biology and computational biology have recently become prominent areas of research in the biomedical community, especially in the area of cell biology. Given that much information on genes and their protein products has become available, the big question is how the individual components interact and work together, and how this determines the functioning of cells, organs, and organisms. Long before the popularity of systems biology in biomedicine, however, such approaches have been used successfully in a different area of biology: population ecology. Research in the area of population dynamics investigated complex interactions between different populations of organisms, such as the dynamics of competition and predation, food webs, community structure, as well as the epidemiology of infectious diseases. In this field, theoretical biology and mathematical modeling have become an integral part of research. Mathematical models allowed people to obtain interesting and counter-intuitive insights into how complex interactions among different populations can play out. Such mathematical studies not only gave rise to interesting theoretical ideas, but also provided the basis for the design of new experimental work and defined major questions and directions of research. Around 1990, such population dynamic concepts, and the use of mathematical/computational approaches, started to be applied to the *in vivo* dynamics between viruses and the immune system. These interactions have many similarities to ecological, epidemiological, and evolutionary principles. Consider the epidemiological spread of a pathogen (such as the common cold) through a population of hosts. Pathogens and parasites reproduce inside infected hosts, are released from the hosts into the environment, and infect other susceptible hosts. Similarly, viruses reproduce inside host cells *in vivo*. Infected host cells produce new virus particles which are released into the extra-cellular environment and which can infect further susceptible host cells. The population dynamic principles are very similar in both cases.

The interactions between pathogens and the immune system is an extremely wide and diverse topic. This book concentrates on a particular branch of the immune system: killer T cells, or cytotoxic T lymphocytes (CTL). This reflects my own research interest and fascination. Killer T cells recognize infected cells, and attack them. Therefore, this book focuses on pathogens which enter host cells. Many pathogens can reproduce within their host without having to infect cells. The main class of pathogens which do infect host cells are viruses. In fact, viruses cannot reproduce without entering the host cells, because the cells' metabolic machinery is required for replication. Hence, the book considers killer T cell responses mostly in the context of viruses.

The purpose of this book is to review how mathematical and computational approaches can be useful to help us understand how killer T cell responses work to fight viral infections. Further, the aim is to demonstrate that such mathematical and computational approaches are most valuable when coupled with experimental work through collaborations. The writing style also reflects this interdisciplinary spirit. While the topic of the book is mathematical modeling, the text is written in a way such that experimental immunologists and virologists should be able to understand the arguments, and to see the biological implications of theory. Experimental readers are encouraged to skip the equations, and to focus on the biological interpretations and discussions.

The work which is reviewed here builds on earlier mathematical research on virus dynamics which considered in detail the relationship between the evolutionary dynamics of viruses *in vivo* and the development of disease in pathogenic human infections such as HIV. Some of this research is summarized in the introductory chapter, and a thorough review is given in a previous book by Martin Nowak and Robert May in the year 2000 (Virus dynamics, Oxford University Press). While the current book aims to cover many important aspects of killer T cell dynamics, it is not intended to address all important issues relevant to the understanding of killer T cell responses. The book concentrates on those topics which have been relatively well worked out, and where theory has been coupled with experimental work in a meaningful way, or where published experimental work provides concrete applications and case studies. It provides a personal view of the subject, guided by my long standing collaborations with various experimental laboratories which work on the biology of killer T cell responses, and which have guided me over the years.

My work, and this book, would not have been possible without the enthusiastic input and guidance from my experimental collaborators. During my PhD years, Paul Klenerman introduced me to LCMV (lymphocytic choriomeningitis virus) infection, and subsequently to Hepatitis C virus (HCV) and cytomegalovirus (CMV) infection. Our interactions continue to inspire my theoretical work. Charles Bangham worked with me on the dynamics of immune responses to human T cell leukemia virus (HTLV-1), and I am extremely grateful to him for his continuous guidance throughout my career. Many of the theoretical concepts in the context of HIV infection came about through my interactions with Jeff Lifson, whose monkeys brought new dimensions to my work. I am grateful to Peter Doherty who shared many valuable immunological insights in the context of murine influenza virus and gamma herpes virus infections. Last but not least, much of my work would not be possible without my collaboration with Allan Thomsen on CTL dynamics in murine viral infections, especially LCMV. His interest in theory, and his openness to apply theoretical thinking to experimental design has not only benefited the quality of the mathematical research, but has also been and continuous to be a tremendous amount of fun.

While my interactions with experimental laboratories has certainly been vital for the development of my work, all this would not have been possible without the guidance of my PhD advisor Martin Nowak, who introduced me to this field and to some of my experimental collaborators, and who opened up many valuable doors and opportunities. Similarly, interactions with theoretical biologists and evolutionary biologists inspired me and influenced my thinking. These include Francisco Ayala, Steve Frank, David Krakauer, Alun Lloyd, and Alan Perelson. Finally, the Department of Ecology and Evolutionary Biology at the University of California Irvine provided a stimulation and productive environment for writing this book.

Irvine, California, USA*Dominik Wodarz*
September 2005

Contents

1 Viruses and Immune Responses: A Dynamical View 1
  1.1 Viruses 3
  1.2 Basic Immunological Background 5
  1.3 Experimental Mouse Models of CTL Dynamics 13
  1.4 Human Pathogenic Infections 15
  1.5 Virus Dynamics and Mathematical Modeling 18
  1.6 Immune Response Dynamics: Structure of the Book 23

2 Models of CTL Responses and Correlates of Virus Control 25
  2.1 Virus Dynamics 26
  2.2 Simplest Model for CTL Dynamics 28
  2.3 Saturated CTL Expansion 31
  2.4 Precursor Versus Effector CTL 32
  2.5 Programmed CTL Proliferation 34
  2.6 Summary 40

3 CTL Memory 41
  3.1 The Generation of CTL Memory: Biological Background ... 43
  3.2 Mathematical Models of CTL Memory 43
  3.3 Resolution of Primary Infection: Mathematical Predictions ... 45
  3.4 Resolution of Primary Infection: Experimental Studies 47
  3.5 Protection Against Rechallenge: Mathematical Predictions ... 49
  3.6 Protection Against Rechallenge: Experimental Studies 51
  3.7 Summary 53

4 CD4 T Cell Help 55
  4.1 Comparison of the Two Helper Pathways 63
  4.2 How Does Help Work? 63
  4.3 Infection Dynamics in Helper Deficient Hosts 66
  4.4 Summary 70

XII Contents

5 Immunodominance 71
  5.1 A Mathematical Model for Multiple CTL Clones 72
  5.2 Theory and Data 75
  5.3 An Unusual Pattern of Immunodominance 77
  5.4 Summary 84

6 Multiple Infections and CTL Dynamics 85
  6.1 Mathematical Model 86
  6.2 Virus Control and Antigenic Heterogeneity 87
  6.3 Two Heterologous Infections 88
  6.4 Multiple Heterologous Infections 90
  6.5 Experimental Studies 93
  6.6 Coinfection: Viruses and Bacteria 94
  6.7 Vaccination 95
  6.8 The Immune Phenome and Aging 96
  6.9 Summary 97

7 Control versus CTL-Induced Pathology 99
  7.1 Basic Mathematical Insights 100
  7.2 CTL-Induced Pathology in LCMV Infection 103
  7.3 CTL-Induced Pathology and HIV Infection 108
  7.4 Summary 111

8 Lytic versus Nonlytic Activity 113
  8.1 Modeling Lytic and Nonlytic CTL Responses 115
  8.2 Effect of Lytic and Nonlytic Immunity on Virus Control 116
  8.3 Noncytopathic Viruses 118
  8.4 More Cytopathic Viruses 121
  8.5 Summary 123

9 Dynamical Interactions between CTL and Antibody Responses 125
  9.1 Modeling Competition between CTL and Antibody Responses 126
  9.2 Competition during Acute Infection 128
  9.3 Effect of Viral Evolution during Chronic Infection 129
  9.4 Application: Experimental Data on HCV Infection 133
  9.5 Summary 136

10 Effector Molecules and CTL Homeostasis 137
  10.1 CTL Homeostasis and Predator–Prey Dynamics 138
  10.2 Effector Molecules and Immunodominance 140
  10.3 The Role of Antigen for CTL Proliferation 142
  10.4 Programmed CTL Proliferation and the Role of CTL Effectors 143
  10.5 Effector Molecules and CTL Homeostasis in VSV Infection ... 144
  10.6 Summary 145

11 Virus-Induced Subversion of CTL Responses 147
- 11.1 A Basic Model for Virus-Induced Impairment of Help 149
- 11.2 What Determines the Outcome of Infection? 151
- 11.3 Robustness of Predictions 153
- 11.4 Experimental Verification: CTL Exhaustion in LCMV Infection 156
- 11.5 Helper-Dependent versus Independent CTL Responses 158
- 11.6 Immune Impairment and the Level of Immune Responses 163
- 11.7 Summary 165

12 Boosting Immunity against Immunosuppressive Infections . 167
- 12.1 Basic Properties of Immune Impairment 168
- 12.2 T Cell Dynamics during Therapy 170
- 12.3 Application: Early Treatment of SIV/HIV 174
- 12.4 Application: Treatment of HCV Infection 178
- 12.5 Treatment Interruptions 179
- 12.6 Summary 181

13 Evolutionary Aspects of Immunity 183
- 13.1 A Single Population of Pathogens 184
- 13.2 Two Competing Pathogen Populations 186
- 13.3 Pathogen Competition and the Evolution of Memory Duration 188
- 13.4 Application to Immunological Observations 191
- 13.5 Summary 192

References 195

Index 217

1

Viruses and Immune Responses: A Dynamical View

One of the most complicated organs of higher organisms is the immune system. The function of the immune system is to fight off pathogenic organisms that enter and grow within the host (for example viruses, bacteria, unicellular eukaryotic parasites such as malaria, and multicellular parasites such as worms). Detailed molecular research has elucidated how immune cells function, that is, how they recognize an invading pathogen and mount orchestrated responses that fight the infection and protect the host. In addition to understanding how the individual components of the immune system work, it is also important to take a "systems approach" and to investigate how the complex interactions between the many components of the immune system work together and determine the outcome of an infection. In a nutshell, this is the subject of this book. In particular, the interactions between pathogens and the immune system can be viewed as an ecological system within the body of an organism. Specifically, the area of population ecology or population dynamics has relevance. Several species of immune cells interact with populations of pathogens in various ways. Two especially important population dynamic interactions that are found in the immune system are predator–prey interactions and competition. (i) When predators capture and kill their prey, they reproduce such that their population size grows. This in turn has a negative impact on the prey population. In the absence of prey, predators die. The outcomes of such interactions can involve cycles in the population sizes of predators and prey that can dampen over time. Similarly, when immune cells encounter pathogens, they divide such that their population size increases, and they remove the pathogens. In the absence of the pathogen, the population of immune cells decays. This can also lead to cycling dynamics that can dampen out over time. (ii) Competition means that two species share a common resource, such as food. The species that can utilize this resource more efficiently will grow better and will be the superior competitor. The superior competitor might drive the inferior competitor extinct (competitive exclusion), or the two species can coexist. Within the immune system, different species of immune cells can potentially recognize the same infection. Each species of cells can expand when it is exposed to this pathogen and kill it. If one immune cell species is more efficient at expanding upon exposure, it fights the infection more efficiently and can prevent another immune cell species from expanding. This is because it reduces the number of pathogens to levels that are too low for the inferior immune cell to become stimulated and divide.

How can such dynamical interactions between different species of immune cells and populations of pathogens be understood? On the one hand, experiments that document the dynamics of immune cells and of pathogens *in vivo* are of central importance. Such data usually show how the populations of pathogens and different immune cells develop over time. In addition, however, a rigorous understanding of such dynamics requires the use of mathematical models that describe and predict the time course of an infection and of immune responses [Levin et al. (1997)]. Mathematical models have been of central importance for understanding the dynamics of populations in an ecological context [Levin et al. (1997)]. In the beginning of the 1990s, such ecological models started to be used by a number of people in order to describe the *in vivo* dynamics between viral infections and immune responses, particularly in the context of human immunodeficiency virus (HIV) infection [Nowak and May (2000)]. Much emphasis was placed on the viral side of these dynamics, including the estimation of basic viral parameters, the evolutionary dynamics of immune escape by the virus, and the analysis of drug treatment in HIV infection. Subsequent work focused on the immune side of these interactions, trying to explain a variety of experimental observations about the dynamics of immune cells in various infections. One particular part of the immune system that is very important in the fight against viral infections are the killer T cells or cytotoxic T lymphocytes (CTL). They basically fight intracellular pathogens. This book will review how mathematical modeling, in close collaboration with experimental research, has contributed to our understanding of the dynamics between killer T cell responses and viral infections.

The aim of this chapter is to provide an introduction to the book. It is written for an interdisciplinary audience. That is, both for mathematical/theoretical biologists and for experimental immunologists who study the dynamics of immune responses. The chapter starts with a brief overview of viral infections and the immune system that is relevant for this book. It subsequently reviews important mathematical work on virus dynamics that is not covered further and that forms the basis for the material presented here. The chapter concludes with an explanation of the structure that underlies the book. The review presented in the current chapter does not aim to provide a full background or a full account of the work in this area of research. Instead, it provides some necessary background that is needed to understand the concepts explored in this book. Mathematically oriented readers who are interested in more detailed explanations of immunological and viral concepts are referred to a standard text book, for example [Pier et al. (2004)]. Biological readers who would like a more thorough introduction to virus dynamics are referred to more detailed reviews such as [Nowak and May (2000); Perelson (2002)]

1.1 Viruses

Among the many pathogen types, this book will concentrate on pathogens that live inside cells (intracellular pathogens). The majority of those are viruses.

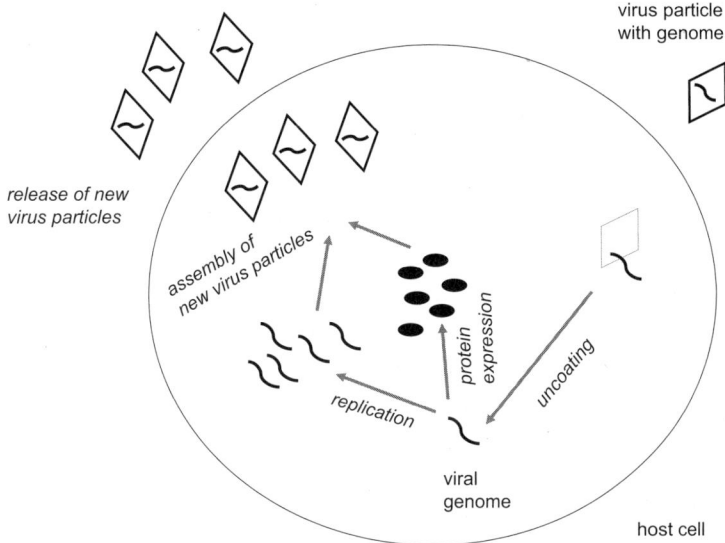

Fig. 1.1. General life cycle of a virus. The virus attaches to the cell via a receptor and enters the cell. There, uncoating occurs where the protein coat is lost and the viral genome is exposed. The viral genome is both replicated and expressed (i.e. viral proteins are made). The newly generated viral proteins associate with viral genomes to build new virus particles. The new virus particles eventually leave the cell.

The life cycle of a virus can have both intracellular and extracellular stages. A virus is basically genetic material wrapped up in a protein coat. It does not have the metabolic machinery to reproduce. For reproduction, the virus has to enter a cell and use the cell's metabolic machinery. In order to enter a cell, the virus usually attaches to some receptor on the cell and is taken in. Then the virus uncoats (Fig 1.1), that is, the protein coat is lost and the viral genome is exposed. Different viruses have different genomes. They can be DNA or RNA, either single stranded or double stranded. The exact mechanism with which reproduction occurs depends on the form of the genome. In addition to reproduction, the viral genome is expressed in the cell. That is, viral proteins are built. The reproduced genomes then associate with the newly generated viral proteins to form new virus particles that eventually leave the cell (Fig 1.1). There are two basic ways in which viruses can exit cells. (i) The progeny viruses can lead to the bursting and death of the cell, whereupon the virus particles are released. This process is called lysis. (ii) The progeny viruses can bud off the cell. That is, they associate with the membrane of the cell and are released, with the cell membrane enveloping the virus particle. The cell stays alive in this case.

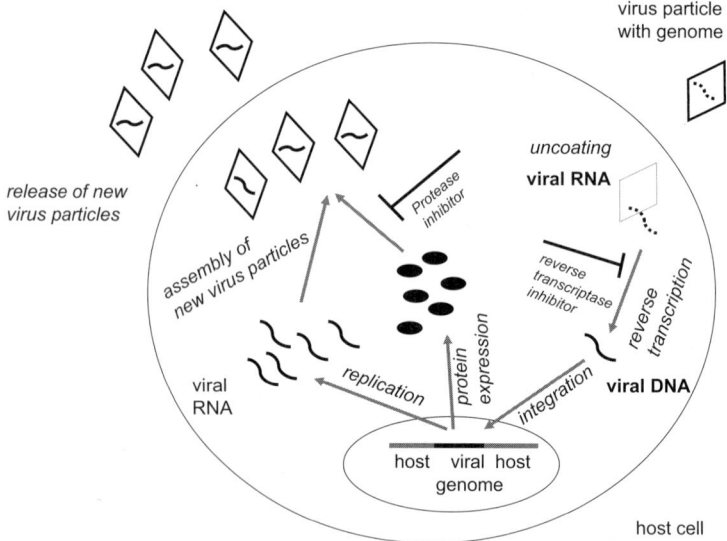

Fig. 1.2. The life cycle of HIV is similar to the general scheme depicted in Fig 1.1, with additional complexity. The viral genome is RNA. It is copied into DNA by the viral enzyme reverse transcriptase. The viral DNA is subsequently integrated into the host genome with the help of the enzyme integrase. Assembly of new virus particles requires the host enzyme protease. Two classes of drugs attack HIV. Reverse transcriptase inhibitors prevent the process of reverse transcription. Protease inhibitors prevent the assembly of new functional virus particles.

Some viruses integrate into the DNA of the host cell during their replication cycle. The best known example of this is human immunodeficiency virus (HIV) (Fig 1.2). Once integrated, the genome of the host cell carries the virus until the cell dies. HIV is an RNA virus. It carries an enzyme called *reverse transcriptase*, which is used to copy the viral RNA genome into DNA (Fig 1.2). This goes against the central dogma by Watson and Crick, which states that genetic information flow can only go from DNA to RNA to protein. Viruses that copy RNA into DNA are called retroviruses. The resulting DNA is then integrated into the host genome with the help of the enzyme *integrase*, which the virus also carries (Fig 1.2). Once integrated, the viral genome is expressed and replicated, resulting in new virus particles that bud off the host cell. The assembly of functional progeny virus requires the enzyme *protease*, which is part of the host cell metabolic machinery (Fig 1.2). Two classes of drugs inhibit the replication cycle of HIV (Fig 1.2). These are reverse transcriptase inhibitors, which prevent the process of reverse transcription, and protease inhibitors, which prevent the assembly of functional virus particles.

1.2 Basic Immunological Background

Immune responses can be subdivided broadly into two categories: (i) Innate or nonspecific responses, and (ii) specific, adaptive responses. Innate immune mechanisms provide a first line of defense against an invading pathogen. They include physical barriers like the skin, changes in environment of the body, such as fever, and immmune cells that can fight pathogens in a nonspecific way. Nonspecific is the key word here and means that these responses cannot specifically recognize the physical structure of the pathogen. Instead they sense that an invader is present and react. While such responses slow down the initial growth of a pathogen, they are usually insufficient to clear an infection. For clearance, a specific and adaptive immune response tends to be required. Specific means that cells bear receptors that can recognize the physical structure of a pathogen. More precisely, they recognize proteins from which the pathogen is built. Upon recognition, these immune cells start to divide and expand. They dramatically increase in number, and this enables them to effectively fight the pathogen, resulting in the resolution of the infection. There is some terminology that is worth to point out here: A substance that is capable of inducing the generation of a specific immune response is called an *antigen*. The site of the antigen that is actually recognized by the receptor of the immune cell is called an *epitope*. The same pathogen can have a variety of epitopes, each of which is recognized by a separate specific immune cell. Therefore, multiple immune cell clones can respond against the same pathogen.

From now on, we will only concentrate on specific, adpative immune responses because those are the ones that are generally necessary for the resolution of infections. The specific immune system has three major branches (Fig 1.3) . Two of them are mainly *effector responses*; that is, they directly fight the pathogen. The third branch is mainly a regulatory response that "helps" the effector responses to become established. The two effector responses are antibodies and CTL. Antibodies are produced by *B cells*. Before an infection has happened, the antibodies tend to be on the surface of the B cells. They serve as the receptor that can specifically recognize the pathogen. When the B cells are exposed to a pathogen, they divide, and secrete the antibodies. The

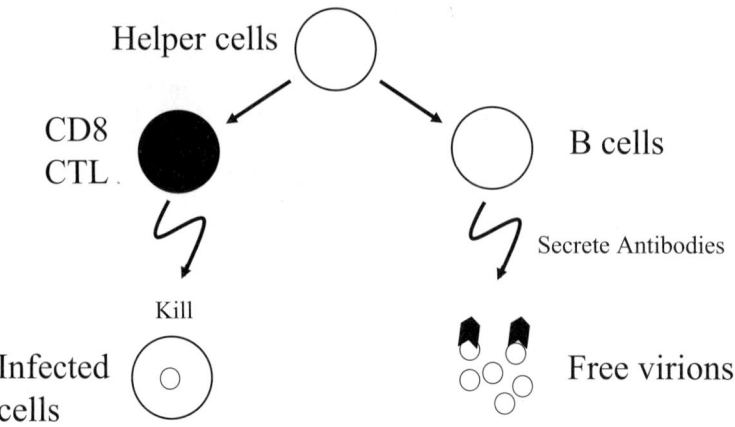

Fig. 1.3. The adaptive immune response consists of the following three main branches. The B cells secret antibodies that neutralize free virus particles. The CTL (also known as CD8 T cells) attack infected cells. The CD4 T helper cells are very important regulators that ensure that CTL and B cell response develop efficiently.

antibodies now float freely, and can attach to the pathogen and neutralize it. Such antibodies are therefore also called *neutralizing antibodies*. For instance, they can neutralize free virus particles. On the other hand, CTL (cytotoxic T lymphocytes or killer T cells), attack infected cells. That is, they fight intracellular pathogens, mostly viruses and some bacteria. CTL bear the T cell receptor on their surface. It can recognize particles of the pathogen that are displayed on the surface of infected cells (see below). When this recognition occurs, the CTL can release substances, and this results in the death of the infected cells. This killing of the infected cells is also known as lysis. In addition, CTL can secrete substances that trigger a reaction inside the infected cells that prevents viral genomes from being expressed. In some cases, this reaction even removes the viral genome from infected cells. That is, besides killing, the CTL can also silence the virus by *nonlytic* means. CTL are also referred to as *CD8 T cells* because they bear the CD8 marker. The regulatory or helping branch of the specific immune system are the so called *CD4 T helper cells*. These are T cells that bear the "CD4 marker", and they help the induction of antibody and CTL responses. They also bear the T cell receptor with which they can recognize the pathogen. They then interact with the CTL and the antibody producing B cells. This interaction makes sure that the CTL and B cells divide and expand to large numbers. In the absence of this CD4 T cell help, CTL and B cells often do not expand properly and the immune response is suboptimal.

This book concentrates on CTL responses. Therefore, the mechanisms with which CTL recognize pathogens is explained in more detail here. Viruses reproduce inside cells. During this process, expression of the viral genome

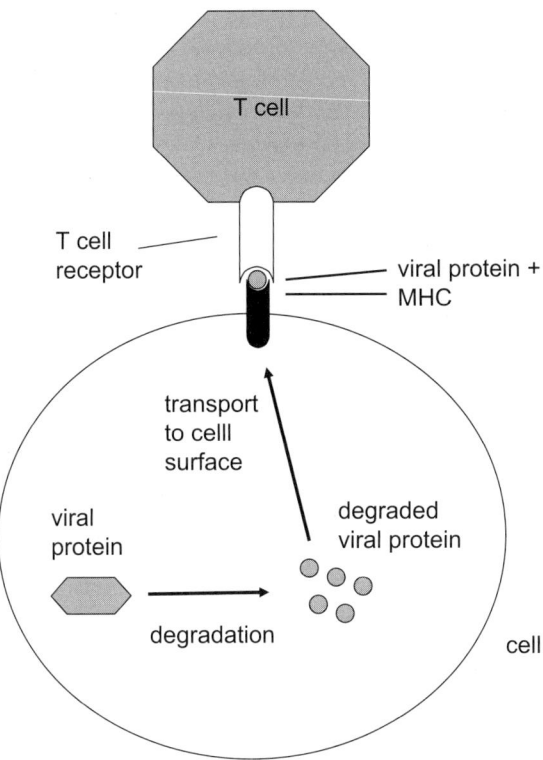

Fig. 1.4. The T cell receptor recognizes viral antigen that is associated with a major histocompatibility complex (MHC) molecule. Inside cells, viral antigen is processed and chopped into small fragments. These associate with MHC molecules and are transported to the cell surface. There are two types of MHC molecules, and this is explained in more detail in Fig 1.5.

produces viral proteins. Some of these proteins are chopped up by specific enzymes inside the cell and are transported to the cell surface. These pieces of viral proteins are then displayed on the cell surface, so that CTL can recognize them. The T cell receptor, however, cannot recognize a piece of viral protein alone. The viral protein associates with a so called *major compatibility complex (MHC) molecule*, and this compound is displayed on the surface of the infected cell and is recognized by the T cell receptor (Fig 1.4). The important point here is that MHC shows a very high degree of polymorphism. Different MHC genotypes can associate with different viral protein segments. Therefore, infected cells from different individuals are likely to display different segments of a given viral protein. As a consequence these individuals differ in the exact immune responses that are induced and thus differ in the susceptibility to the infection. There are two basic types of MHC molecules: MHC class I and MHC class II. CTL recognize antigen that is associated with MHC class I. Every cell of the body produces MHC class I molecules. On the other hand, antigen that is associated with MHC class II molecules is recognized by CD4 T helper cells. Antigen in conjunction with MHC class II is not displayed on infected cells. Instead, it is displayed on so called *antigen presenting cells (APCs)*. Examples are dendritic cells and macrophages. These are not infected by the virus, but they eat up virus particles. The captured virus particles are processed by the APCs, and the resulting protein segments are displayed on the surface of the APCs in conjunction with MHC class II. The purpose of APCs is to capture pathogens and to display them to specific immune cells in order to start up a response.

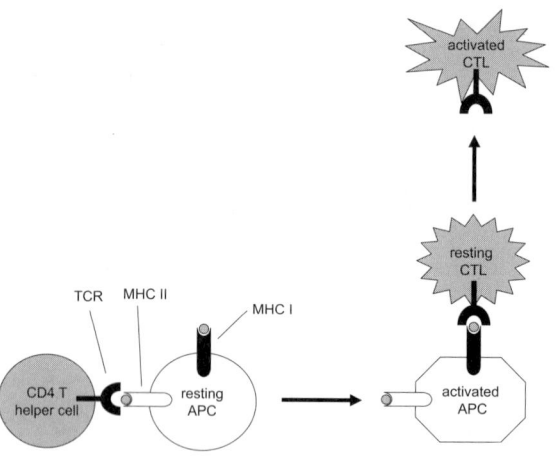

Fig. 1.5. Pathway of CTL activation. There are two types of MHC molecules. MHC I is found on every cell in the body. Upon infection, the virus inside the cell is chopped up and the fragments are transported to the surface for display with MHC I. CTL recognize antigen + MHC I. MHC II is found mostly on antigen presenting cells (APCs). They eat up the antigen and display fragments of the virus on their surface together with MHC II. This is recognized by CD4 T helper cells. Since every cell in the body bears MHC I, APCs can contain both types of MHC molecules. First, APCs interact with CD4 T cells through antigen/MHC II recognition. This activates the APCs, and also induces the CD4 T helper cells to produce certain signaling molecules. The activated APCs interact with the CTL via antigen/MHC I recognition. This activates the CTL and starts a response.

The overall picture of how CTL, CD4 T helper cells, and APCs interact to give rise to a successful CTL response is as follows [Ridge et al. (1998); Schoenberger et al. (1998)] (Fig 1.5). Immune cells with a specificity for a virus are present in the host at very low numbers if the host has never been infected by this virus before. The host is said to be naive. The infection occurs in some specific tissue, such as the lung, liver, etc. The APCs eat up the virus particles at this site and bring them to the *lymph nodes* that are locations where most immune cells are parked. The APCs express both MHC class I and MHC class II. This allows a set of reactions to occur in the lymph nodes. The CD4 T helper cells recognize antigen displayed in conjunction with MHC class II. The interaction between APCs and the CD4 T cells activates the APCs. In addition, it triggers the CD4 T cells to divide and to produce certain signaling chemicals called *cytokines*. The activated APCs also display the viral proteins in conjunction with MHC class I, and can therefore interact specifically with the CTL. This activates the CTL and induces them to divide such that the population of these cells grows to high numbers (Fig 1.6). Since one cell gives rise to a large number of cells through division, this is also referred to as *clonal expansion*. CTL division is further supported by the cytokines that are secreted from the CD4 T helper cells. The activated and expanding population of CTL and CD4 T cells migrates away from the lymph nodes towards the anatomical site of infection. There, the CTL recognize pieces of viral proteins displayed on the infected cells themselves in conjunction with MHC class I molecules. This interaction induces the CTL to perform antiviral activity. As already mentioned briefly earlier, there are several ways in which CTL can perform antiviral activity [Kagi and Hengartner (1996); Kagi et al. (1995a); Kagi et al. (1996); Kagi et al. (1995b)]. On the one hand, CTL can kill, or lyse, infected cells. The main molecule that mediates lysis is perforin, which is secreted by CTL and induces programmed cell death (apoptosis) in the infected cell. A less important mechanism of lysis is the interaction between FAS and FAS ligand when CTL recognize an infected cell. On the other hand, CTL can have nonlytic antiviral activity. They secrete molecules that either silence viral gene expression in infected cells, remove the genome from the infected cells, or prevent the virus from infecting susceptible cells. Interferon gamma (IFN-γ) is an example of a molecule that is secreted from CTL and that can inhibit viral replication inside infected cells.

As the CTL perform antiviral activity, virus load declines and clearance of the virus from the host occurs in many cases. As the infection is resolved, the population of CTL declines (Fig 1.6). This is often referred to as the contraction phase. However, it does not decline to the same low levels from where the response started. Instead it settles around an elevated level, and the CTL persist at this elevated level for long periods of time in the absence of any further exposure to the virus (Fig 1.6). This is called immunological memory and is observed in all branches of the specific, adaptive immune system. If a heightened number of immune cells remains after the resolution of infection, it is thought that the host can react more efficiently if it is reinfected with the same pathogen again. Such a secondary infection will not result in much virus growth and the host is protected from symptoms and disease. Immunological memory is also the basis that underlies the protective function of vaccines.

As stated above, in many cases, the expansion of a CTL response and of immune responses in general results in the clearance of the virus from the

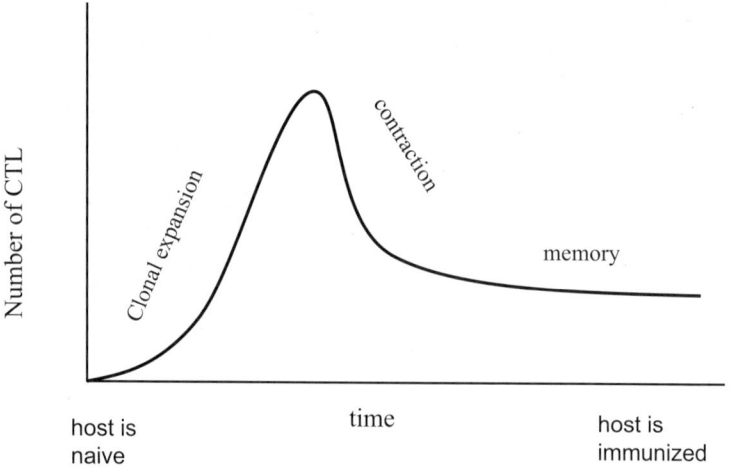

Fig. 1.6. Before a host is exposed to antigen, there are few specific CTL and specific immune cells in general. The host is said to be naive. Upon exposure to the virus, the CTL divide and expand (clonal expansion). They fight the virus population, and as the infection is resolved, the number of CTL declines to a certain degree. It subsequently settles around an elevated level in the long-term. This is called *immunological memory*. The elevated number of memory CTL can fight the virus more efficiently if it infects the host a second time.

host. However, viruses can fight back against the immune system, and this can result in failure to clear and thus in persistent infection. Various mechanisms can lead to this effect, and many of them are present with HIV. Some examples are listed as follows. As explained above, specific immune responses can recognize a defined part of a viral antigen (i.e. a viral epitope). The T cell receptor is specific for a given epitope. If the virus mutates this epitope, the T cell will not be able to recognize it anymore. This renders the T cell useless, because it cannot recognize or fight the virus. This is called *antigenic escape* (Fig 1.7) and has been documented extensively with HIV, and hepatitis B and C virus, among other infections. Another important mechanism by which viruses can fight back against the immunity is to impair the central component of the specific immune system: the CD4 T cells (Fig 1.8). If the CD4 T cell response is impaired, then the CTL and antibody responses cannot be mounted effectively, resulting in weak responses. This in turn can result in failure to clear the infection. HIV is a prime example of this. The CD4 T cells are a main target cell that HIV infects. Infection results in CD4 T cell death. Thus, no CD4 T cell help is available for HIV specific CTL or antibody responses, and this compromises the ability of the immune system to fight the infection. Other infections impair the CD4 T cell response by different mechanisms, without killing the helper cells. An example is hepatitis C virus (HCV)

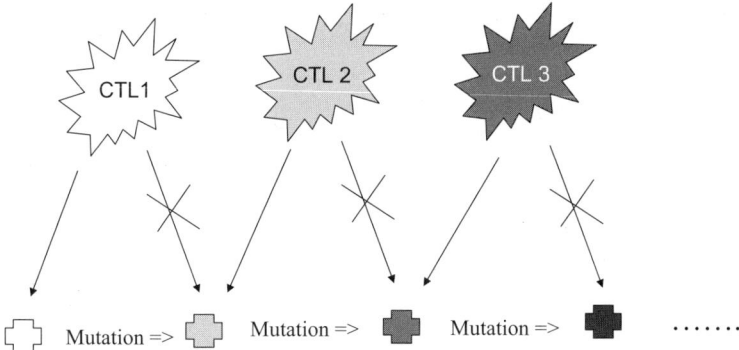

Fig. 1.7. The concept of antigenic escape. One CTL is specific for a certain antigen. If it mutates, the CTL cannot recognize it anymore. Another CTL response may arise that can recognize the new mutant. If this mutant mutates further, it can also escape the new response. Thus, there can be a race in which the CTL response tries to adapt to an evolving virus, and the evolving virus continuously escapes the CTL.

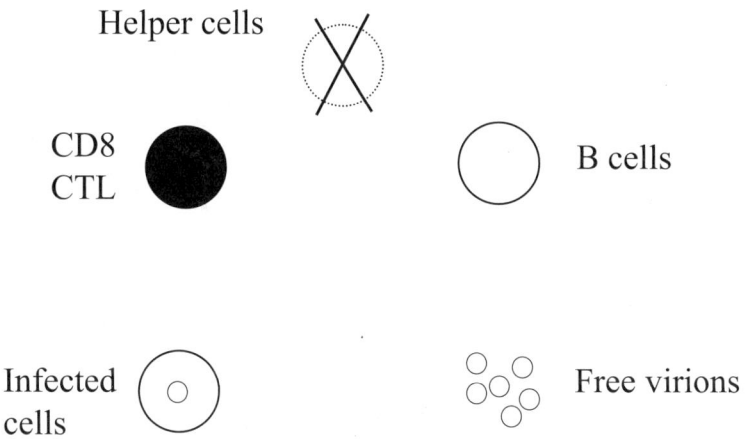

Fig. 1.8. A very effective way for viruses to fight specific immune responses is to impair CD4 T cell help. Since CD4 T helper cells are central regulators of both B cell and CTL responses, these effector responses cannot fight the virus efficiently anymore in the absence of CD4 T cell help.

infection. The mechanism underlying this impairment is unclear. Some viruses can impair T cell responses without killing them when not all necessary activation signals are delivered upon stimulation. In this case, T cells tend to remain silent, a state called *anergy*. Finally, another way to achieve persistent infection is to simply hide from the immune system. In other words, the virus can establish a latent infection. The virus infects cells, but does not produce progeny virus for extended periods of time. As a result, no viral antigen is displayed on the cell surface, and the virus is invisible to the immune system. Such latent phases are followed by "lytic phases" where cells "wake up" and progeny virus is produced and spreads to other cells. The immune response will prevent the rise of the virus to high levels, but will fail to remove the virus because of the continued presence of latently infected cells.

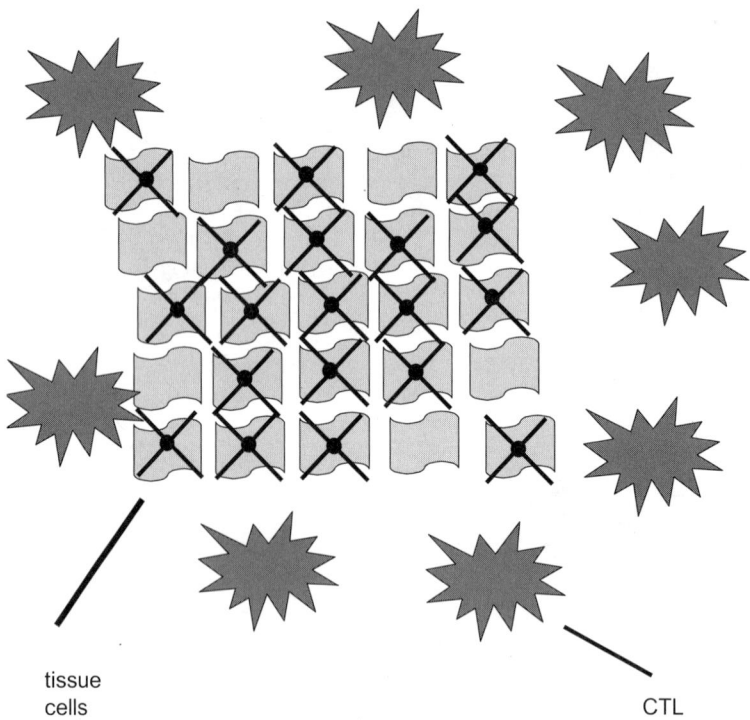

Fig. 1.9. The concept of CTL induced pathology. Infected tissue cells are marked in the diagram. If many tissue cells are infected by the virus, and a specific CTL response persists continuously, many tissue cells can be killed by the CTL, and this can result in pathology.

Most of the time, CTL responses are beneficial for the host. They expand, fight the virus, and eliminate it. There is, however, also a negative side. If the CTL response is strong enough to expand and kill infected cells, but not strong enough to eradicate the infection from the host, it is possible to observe high levels of persistent virus replication combined with an ongoing CTL response. If the virus persists in many cells of a tissue, the ongoing killing by the CTL can severely damage the tissue. This can lead to the death of the host. This concept is called CTL-induced pathology [Thomsen et al. (2000); Zinkernagel (1993)] (Fig 1.9). This is an important concept that will be explored in this book.

1.3 Experimental Mouse Models of CTL Dynamics

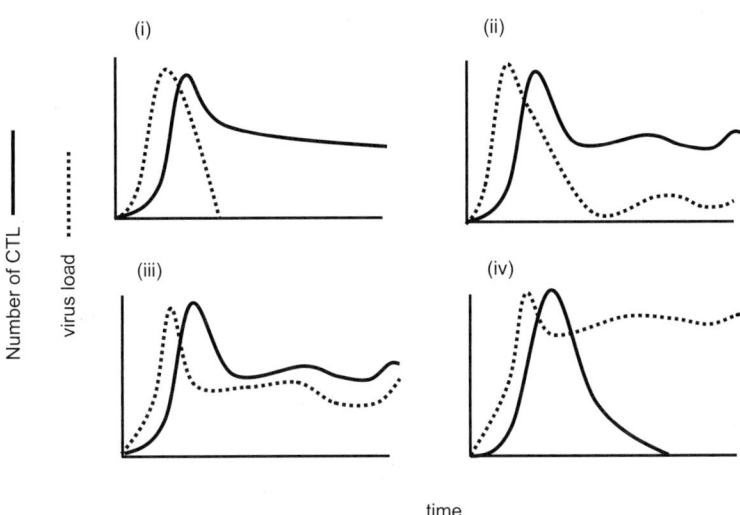

Fig. 1.10. Different outcomes of LCMV infection, depicted as schematic time series. (i) The CTL can clear the virus. (ii) If the CTL fail to clear the virus, they can control the virus in the long-term at low levels. No pathology is observed. (iii) If the virus persists at higher levels, and if the CTL response also persists, then the killing of infected tissue cells can result in CTL-induced pathology. (iv) Finally, the CTL response can go extinct, and the virus replicates at high levels. This also called CTL exhaustion. Because the virus is noncytopathic (does not kill infected cells), no pathology is observed.

The dynamics of CTL responses to viral infections are studied experimentally by monitoring the number of CTL and the number of virus particles or infected cells over time. This requires detailed data. For human pathogens, such data are difficult to obtain, although a wealth of data has accumulated from HIV infected patients. A more controlled and accurate way to study CTL dynamics *in vivo*, however, is to use mouse models of virus infections. Mice can be infected in a controlled way, and the populations of immune cells and viruses can be measured at regular intervals. In addition, the mice can be modified genetically by deleting certain functional parts of the immune system. Such mice are referred to as *knockout mice*. For example, mice can be engineered to lack CD4 T cell help, or to lack CTL-secreted perforin that is required to kill infected cells.

A particularly well studied mouse infection is lymphocytic choriomeningitis virus (LCMV)[Lehmann-Grube (1971)]. It is a noncytopathic, or noncytotoxic virus. That is, infection does not result in the death of the host cells. Several LCMV strains exist that differ in their replication rate. LCMV infection can result in a variety of outcomes, depending on the viral strain, the infectious dose, and the genotype of the mice [Moskophidis et al. (1995a); Thomsen et al. (2000)]. The following outcomes can be observed (Fig 1.10).

(i) CTL-mediated clearance of the virus from the host. The CTL expand, remove the virus, and settle around an elevated memory level.

(ii) CTL-mediated long-term control of the infection. This outcome is similar to the first scenario, except the CTL do not manage to clear the virus. Instead the virus persists at low levels. Persistent virus replication drives an ongiong CTL response that keeps the infection at bay. No pathology is observed in mice.

(iii) CTL-induced pathology. This occurs if the CTL do not clear the infection and also fail to keep the persisting virus population at low levels. Now, there is a large number of infected cells, in combination with an ongoing CTL response that can kill these infected cells. Consequently, many tissue cells die, and the mice waste away and die.

(iv) CTL exhaustion. This means that although the CTL response initially expands, it later decays and goes extinct. No memory cells are generated. As a result, the virus persists at high levels. Because LCMV is noncytotoxic, the mice remain healthy.

As the rate of viral replication is increased, or if the initial infectious dose is increased, the infection dynamics shift from outcome (i) to outcome (iv) [Moskophidis et al. (1993b)]. A faster initial virus spread in the host weakens the ability of the CTL to catch up with the virus. Host parameters, most importantly the strength of the CTL response, also have an influence. Mice deficient in CD4 T cell help can only mount compromised CTL responses, and this pushes the outcome towards CTL-induced pathology and CTL exhaustion [Christensen et al. (2001)].

Studying mouse infections not only gives interesting insights about factors that can influence the ability of CTL to successfully resolve a virus. It also has implications for important human pathogens, such as HIV. A central feature of HIV infection is that it infects and kills the CD4 T helper cells. Therefore, the analysis of infection dynamics in mice that are deficient in CD4 T cell help could shine light onto the factors that govern HIV dynamics. Similarly, the analysis of CTL-induced pathology in mice has implications for hepatitis B and C virus infections. In these scenarios, a large number of liver cells become infected, and CTL-mediated killing of liver cells could contribute to liver pathology.

While LCMV is the most widely studied mouse virus, other mouse infections are also used to study CTL dynamics *in vivo*. These include vesicular stomatitis virus (VSV), influenza virus, and gamma herpes virus. The latter results in the generation of latently infected cells. This prevents immune responses from eradicating the virus from the host. Once infected, the latently infected cells can periodically "'wake up"' and become active. The best case scenario is that an efficient, ongoing CTL response controls the infection at low levels in the long-term.

1.4 Human Pathogenic Infections

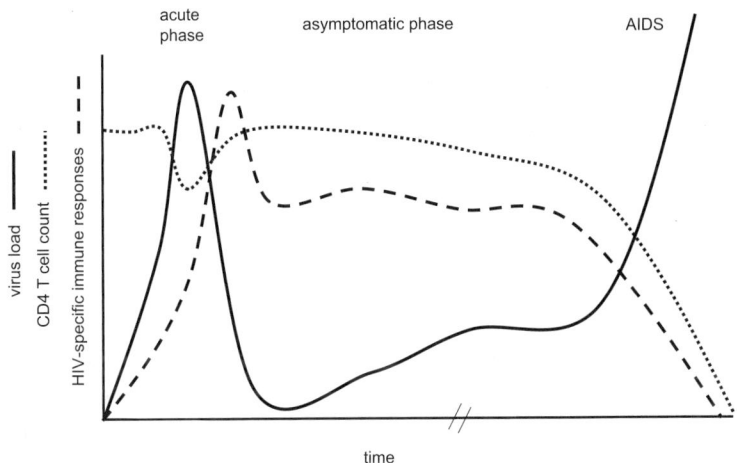

Fig. 1.11. Phases of HIV infection. During the acute phase, virus load rises to high levels. Acute symptoms are experienced. The rise of virus load correlates with a temporary dip in the CD4 T cell count. Immune responses rise, and virus load is reduced. This marks the beginning of the asymptomatic period. During this period, virus load only rises slowly, and the CD4 T cell count and immune responses stay relatively stable. After a variable period of time, virus load rises more sharply, and the CD4 T cell count drops sharply. This marks the beginning of AIDS. AIDS is defined if the number of CD4 T cells has dropped from the normal level (1000 cells / μl blood) to 200 cells / μl blood. At this end stage of the disease, the HIV specific immune responses also disappear.

Work on mouse models of viral infections provides important information about the dynamics of CTL responses, and these insights help in the understanding of human pathogenic infections, which are characterized by a considerably higher degree of complexity. The best known human viral pathogen is human immunodeficiency virus (HIV). HIV infection is characterized by three phases (Fig 1.11). In the earliest stages of infection, after the virus has entered the host, we observe the acute phase. The virus grows to high levels and immune responses rise. The immune responses reduce virus load to lower levels, but fail to clear the virus. The CD4 T cell count takes a temporary dip before returning to normal levels (1000 cells / μl blood). During this phase, infected individuals can experience symptoms typical of viral infections in general, such as fever, rash, and fatigue. Once virus load has fallen to lower levels, these symptoms subside, and this marks the beginning of the asymptomatic or chronic phase of the infection. During this phase, virus load remains relatively low and the CD4 cell count remains relatively high. The final stage of the infection is the development of AIDS. This is characterized by a fall of the CD4 T cell count to less than 200 cells / μl blood, and a sharp rise in virus load. Because the body has a highly reduced CD4 T cell count, the immune system does not function anymore, and the patient dies from a variety of infections that would otherwise be cleared. Such infections are called opportunistic infections. The duration of the asymptomatic phase of infection is highly variable. On average, it takes between 5-10 years. However, some patients develop AIDS rapidly after only a few months, while so called *long-term nonprogressors* do not develop any signs of AIDS for as long as 15-20 years after infection. Such patients are characterized by very low virus loads and high levels of immune responses. The reasons for the transition from the asymptomatic phase to the development of AIDS are unknown.

A central feature of HIV is that it infects and kills CD4 T helper cells. As explained above, these are the central regulatory branch of the specific immune system that enable CTL and B cell responses to develop fully. The virus can infect the entire CD4 T cell population. This includes CD4 T cells that are specific to a wide variety of pathogens. This accounts for the general subversion of immunity observed in AIDS patients. It also includes the population of CD4 T cells that are specific to HIV itself. In fact it has been found that HIV might preferentially infect HIV specific CD4 T cells [Douek et al. (2002)]. Therefore, it attacks the immune responses against itself and this can contribute to the ability of the virus to persist in the host. Immune responses against HIV typically rise during acute infection and persist during the asymptomatic phase. They tend to collapse as AIDS develops (Fig 1.11). This applies both to neutralizing antibody responses and CTL responses. It is thought, however, that these responses are compromised already early on during the acute phase of infection [Rosenberg et al. (1997); Rosenberg et al. (1999)].

Another important feature of HIV is its high mutation rate of about 10^{-4}-10^{-3} [Nowak (1990)]. This is because reverse transcriptase lacks proof reading mechanisms. During genome copying or replication, errors are incorporated. In DNA replication, such errors are corrected by proof reading mechanisms. When viral RNA is copied into DNA, such errors are not corrected, and this results in an elevated mutation rate. It has been estimated that the HIV mutation rate optimizes the ability of the virus to escape from immune responses [Nowak (1990)]. More generally, it allows the virus to have an enormous evolutionary potential. Because of the high mutation rate, the virus population consists of many subtypes that all differ slightly from each other, but have a similar core or master sequence. Such a collection of related types is called quasispecies.

As mentioned briefly earlier in this chapter, two types of drugs are used to attack the life cycle of HIV. (i) Reverse transcriptase inhibitors prevent the process of reverse transcription; that is the process of copying viral genomic RNA into DNA. If reverse transcription does not occur, the virus cannot integrate into the host genome and the cell cannot become infected. (ii) Protease inhibitors prevent the assembly of new functional virus particles in the cells. This also inhibits the spread of the virus to further susceptible target cells. The net effect of both classes of drugs is the same. They prevent the virus from spreading to further host cells. Cells that are already infected, however, remain infected and alive in the presence of drug therapy. Typical therapy regimes involve a combination of three drugs, such as two different reverse transcriptase inhibitors and one protease inhibitor. This minimizes the chances that drug therapy fails as a result of drug resistance. This regimen is referred to as combination therapy or highly active antiretroviral therapy (HAART).

HIV is probably the most complex infection with regards to the dynamical interactions between viruses and the immune system. Another human disease considered in this book is hepatitis, especially hepatitis C virus (HCV) infection. During the acute phase, a relatively small fraction of patients clear the virus from the blood, and this is associated with very strong and sustained CTL responses [Lechner et al. (2000a); Lechner et al. (2000b); Lechner et al. (2000c)]. Most patients, however, fail to clear the virus and develop a productive, persistent infection, associated with weaker and less sustained CTL responses. HCV infects liver cells and causes liver pathology. Pathology does not, however, occur immediately. The virus can replicate in the host for up to 20 years, before symptomatic infection is observed. The reason for this is not known. While fundamentally different from HIV infection, HCV shares the feature that specific CD4 T cell responses to itself are compromised [Barnes et al. (2002)]. This might contribute to the ability of the virus to persist in the host. In contrast to HIV, however, HCV does not infect and destroy CD4 T cells. The mechanism that underlies the impairment of the HCV specific CD4 T helper cell responses remains unclear. Since CD4 T helper cell responses are impaired in HCV infection, however, it provides a nice comparison to HIV.

1.5 Virus Dynamics and Mathematical Modeling

Mathematical models provide an essential tool that complements experimental observation in the study of CTL dynamics. The complex and nonlinear nature of the interactions that occur during the generation of immune responses renders a rigorous understanding of the outcome of infection difficult to achieve by verbal arguments alone. Mathematical models go beyond verbal or graphical reasoning and provide a solid framework that captures a defined set of assumptions and follows them to their precise logical conclusions. This framework can be used to generate new insights, to create hypotheses, and to design new experiments. Such mathematical models of immune responses are the subject of this book. Earlier work that is not covered in this book concentrates more on the viral than on the immune side of these interactions, and provides a basis for the material covered in the current volume. Much of this work is concerned with HIV infection because of the amount of data available. It is reviewed briefly as follows.

The life cycle of viruses, such as HIV, can be modeled by basic infection dynamics equations. These are reviewed in detail in Chapter 2. The equations describe the development of the populations of uninfected and infected cells, as well as free viruses over time. Since drugs against HIV infection basically inhibit the spread of the virus to new cells, but leave infected cells intact, therapy is modeled by a reduction in the overall replication rate of the virus. Upon initiation of treatment, the infected cells decay with a certain death rate (natural + virus-induced death), and no newly infected cells are generated. During chronic HIV infection, virus load fluctuates around relatively stable levels. Upon drug treatment, the virus load of patients drops exponentially. With the help of such kinetic data, mathematical models have allowed researchers to calculate the death rate and thus the half-life of infected cells (Fig 1.12) [Ho et al. (1995); Perelson et al. (1997); Perelson et al. (1996); Wei et al. (1995)]. This is an indicator of the turnover rate of the virus. The faster the death rate of infected cells (and thus the shorter their half life), the faster the rate of viral replication has to be in order to maintain virus load at the relatively stable levels observed in chronic infection. These studies gave rise to very important insights. Before, it was thought that during the chronic phase of infection, the virus is latent. That is, cells do not express the viral genome. Consequently, AIDS is triggered by some event that "'wakes up"' the virus, leading to productive virus replication and pathology. This turned out not to be true (Fig 1.12). The half life of infected cells was found to be relatively short, on the order of 1-2 days (Fig 1.12) [Ho et al. (1995); Perelson et al. (1997); Perelson et al. (1996); Wei et al. (1995)]. Free virus particles were found to have an even shorter half-life of about 6 hours. If infected cells and free virus particles die so often, they need to be produced with a fast rate in order to account for the constant virus load. That is, HIV has a high turnover rate in the asymptomatic, chronic phase of the infection. Therefore, the infection is not latent during this period. This has far reaching consequences.

A high turnover rate means that many replication events occur during the asymptomatic phase. Therefore, many mutants can be generated that escape immune responses or antiviral drugs.

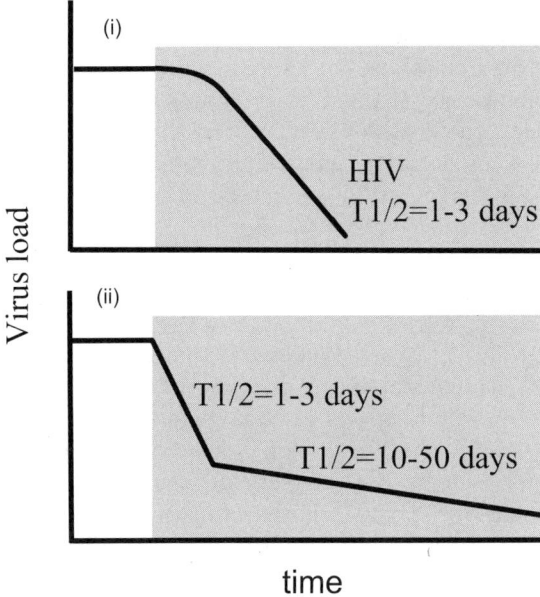

Fig. 1.12. HIV dynamics upon drug therapy during asymptomatic HIV infection. (i) First insights. Upon start of therapy, virus load declines exponentially (i.e. a straight line on a log scale). According to mathematical models, the slope of this line provides an estimate of the death rate of infected cells, and this allows us to calculate the half-life of infected cells. The big result was that the half-life of infected cells is short, on the order of 1-3 days. This means that the viral turnover rate is high during the asymptomatic phase of the infection. This decay represents the decay of infected CD4 T cells. (ii) Later results showed that viral decline upon drug therapy is in fact biphasic. The fast rate of decline is followed by a much slower rate of decline, where the half-life of infected cells is around 10-50 days. This represents the decline of infected antigen presenting cells such as dendritic cells and macrophages. As the viral decline continues it can slow down further because of the presence of latently infected cells.

These results also provided optimism for the treatment of HIV infection. If the infected cells decay with such a fast rate upon drug therapy, then a duration of therapy of about three years would be sufficient to eradicate the virus from the host. Further studies that were conducted over a longer period of time and with a combination of drugs, however, revealed that the viral decay upon drug therapy was biphasic (Fig 1.12) [Perelson et al. (1997)]. During the first phase, the half life of infected cells was around 1-2 days, as explained above. Subsequently, however, there is a second and much slower phase of virus decay. During this phase, the half-life of infected cells is of the order of 10-50 days (Fig 1.12). As time goes on, the rate of virus decay can slow down further, with half-lives of infected cells reaching approximately 100 days. Therefore, in order to eradicate HIV from patients, drug therapy would have to be applied for a duration of time that is longer than the life span of patients. These different phases of virus decay are explained by the fact that HIV can infect a variety of different cell types. The first and fast phase of virus decay reflects the death of infected CD4 T helper cells. The second and longer phase of virus decay reflects the death of infected antigen presenting cells. HIV can infect a variety of such cells, most importantly dendritic cells and macrophages. The slowdown to even lower rates of virus decay reflects the presence of truly latently infected cells that do not actively express the HIV genome [Chun et al. (1997)].

The finding that the virus turns over with a high rate during the asymptomatic phase of the infection means that the virus has an enormous potential to evolve. Two important aspects of viral evolution is the emergence of drug resistant virus strains, and the rise of virus strains that escape immune responses (antigenic escape). Both aspects are reviewed as follows.

HIV can acquire resistance to antiviral drugs by simple point mutations. Given the high mutation rate of the virus, drug resistance is generated easily [Richman (1994); Richman (1996); Richman (1998)]. Mathematical models have been used to work out the principles according to which drug resistant virus strains emerge, and how treatment failure as a result of resistance can be avoided. One important question is when exactly drug resistance evolves. Does this occur during the phase of therapy, or in the pretreatment phase of the infection [Bonhoeffer et al. (1997); Bonhoeffer and Nowak (1997); Ribeiro and Bonhoeffer (2000)]? Mathematical models predict that the chances of resistance evolving during treatment are very small compared to the chances of resistance evolving in the pretreatment phase [Bonhoeffer and Nowak (1997); Ribeiro and Bonhoeffer (2000)]. Therefore, if treatment fails as a result of resistance, it is very likely that resistance evolved before the start of therapy. Therefore, starting therapy as early as possible in the disease process would minimize the chances that resistant viruses exist when the drug regimen is started, and that therapy fails. The first treatment regimes for HIV infected patients involved the use of a single reverse transcriptase inhibitor. This treatment was not successful because resistant mutants always emerged. Now several drugs with different mechanisms of action are available, and they can be used in combination. How many drugs should be used in combination in order to prevent the emergence of a mutant virus that is resistant to all drugs in use? Mathematical models suggest that a combination of three drugs should be used [Ribeiro et al. (1998)]. If a single drug is used, it is very likely that at least one resistant mutant exists upon initiation of treatment. If two drugs are used, this likelihood is lowered but still relatively large. However, if three drugs are used in combination, then it is extremely unlikely that a mutant that is resistant against all drugs in use is found upon start of therapy. These calculations take into account the mutation rate of HIV, and a fitness cost of resistant mutants (approximately 10%). They further assume that the drugs affect different targets and that a mutant resistant against one drug is not resistant against any of the other drugs in use. These are realistic assumptions, and clinical data indicate that a combination of three drugs results in sustained suppression of virus load to undetectable levels. This requires that the drug schedule is strictly maintained. If patients do not adhere to their drug schedule, these calculations break down. In particular the chances that a resistant mutant is generated during therapy increases sharply [Wahl and Nowak (2000)]. This is because in this case, the virus is only suppressed partially by the drugs, which allows for higher levels of viral replication during therapy.

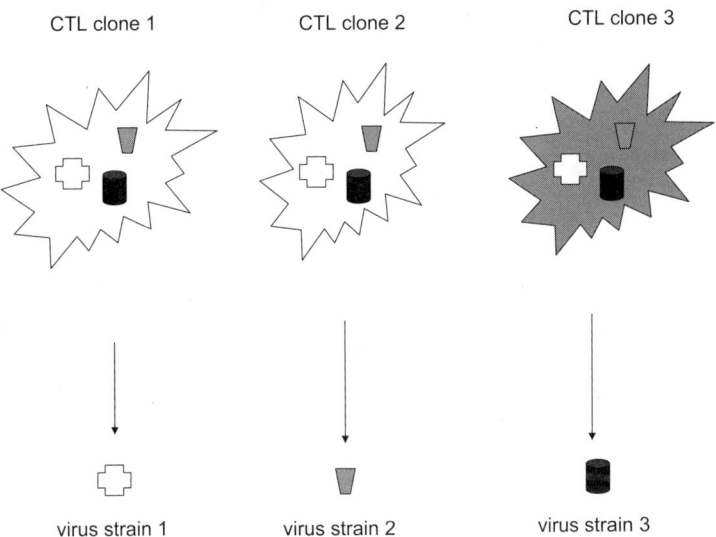

Fig. 1.13. Asymmetry in the fight between HIV and the immune system. While a given immune cell clone can only recognize the one viral strain it is specific for, all viral strains can infect all susceptible CD4 T cells regardless of their specificity.

The other area in which viral evolution might play a major role is the emergence of mutants that escape immune responses (antigenic escape mutants). The evolutionary dynamics of antigenic escape have been explored extensively with mathematical models [de Boer and Boerlijst (1994); Nowak (1992); Nowak (1996); Nowak et al. (1991); Nowak and May (1991); Nowak et al. (1995a); Nowak et al. (1995b); Regoes et al. (1998)]. It is possible that antigenic escape contributes to the ability of the virus to persist in the host and to evade clearance. While the immune response fights the infection and reduces virus load, an escape mutant emerges and grows. A new immune response can be generated against this escape mutant. However, before it is cleared, this mutant can acquire another mutation that enables it to also escape from the new response. Consequently, the mutant will grow. Therefore, while the immune system can adapt and mount new responses against variant viruses, the virus population continues to evolve away from these responses and to avoid clearance. Whether antigenic escape is an important mechanism that contributes to viral persistence is unclear. Defects in certain branches of immune responses (such as the impairment of CD4 T cell help) might be a more important determinant of viral persistence [Rosenberg et al. (1999)]. Escape mutants have been identified clearly in a variety of cases, both in the context of antibody and CTL responses [Goulder et al. (1997a); Goulder et al. (1997b); Goulder et al. (1997c); McAdam et al. (1995); McMichael et al. (1996); McMichael et al. (1995); McMichael and Phillips (1997); Phillips et al. (1991); Price et al. (1997a); Price et al. (1997b); Price et al. (1999); Saag et al. (1988); Wei et al. (2003)]. Several studies indicate that the emergence of immune escape mutants might correlate with advanced progression of the disease and the development of AIDS. Mathematical models suggest that immune escape might indeed drive progression of the disease. A particular study has defined an antigenic diversity threshold, beyond which the disease status changes from being asymptomatic to the development of AIDS [Nowak et al. (1991)]. The argument is as follows. There is an asymmetry in the way in which the immune system fights the virus, and the virus fights the immune system (Fig 1.13). As explained in the beginning of this chapter, immune responses are characterized by specificity. That is, a response against the wild-type does not recognize a mutant, and a response against a mutant does not recognize the wildtype, or any other mutant. On the other hand, any HIV strain/mutant can infect any susceptible CD4 T cell regardless of its specificity (Fig 1.13). Therefore, as long as the number of HIV strains remains below a threshold, the immune system can keep up with virus evolution and suppress the virus to relatively low levels. However, once the number of viral escape mutants has crossed a threshold, the immune responses collapse and the virus grows to high levels (Fig 1.14). If there are too many viral mutants, a given CD4 T cell clone can only recognize and fight a small fraction of the entire HIV population in the host (the virus strain to which it is specific), but all the virus strains can infected and kill all CD4 T cells. Thus, in the presence of a large number of escape mutants, the immune system loses this fight and becomes overwhelmed (Fig 1.14). Whether evolution of immune escape really drives the progression of the disease towards AIDS is not clear. Data from HIV infected patients argue both in favor and against this theory [Goulder et al. (1997c); McMichael and Phillips (1997); Price et al. (1997a); Wolinsky et al. (1996)]. Further interesting data come from monkeys who are naturally infected with simian immunodeficiency virus (SIV, the monkey equivalent of HIV), and do not develop any disease. In this case, it is possible that the virus persists at high levels, evolves towards high antigenic diversity, yet never develops any disease [Broussard et al. (2001); Silvestri et al. (2005)]. This observation suggests that other mechanisms must be necessary for the development of AIDS. However, whether evolution to increased amounts of viral diversity is the key to HIV disease progression or not, such mathematical modeling approaches give rise to the more general message that the evolution of the virus *in vivo* might be required for the development of AIDS (Fig 1.14). The traits of the virus can change, which can switch the infection from an asymptomatic to a pathogenic state. For example, the virus can evolve to replicate with a faster rate [Tersmette et al. (1989)], use diffcrent coreceptors [Moore et al. (1997); van't Wout et al. (1994)], or change its toxicity for its target cells [Rudensey et al. (1995)]. More studies are required to determine the key factors that drive HIV disease progression.

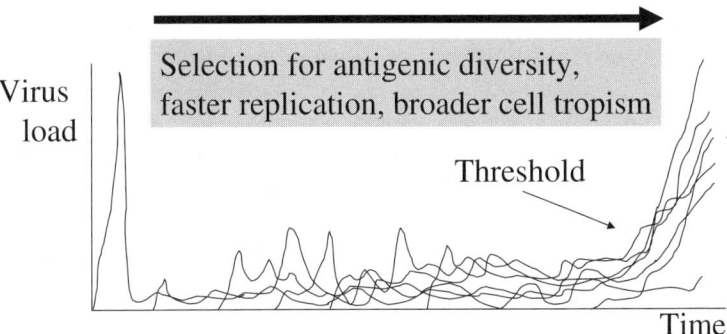

Fig. 1.14. Viral evolution of antigenic diversity can in theory drive the progression to AIDS. As more and more virus strains escape immune responses and accumulate, a given immune cell clone can recognize smaller fractions of the viral population, while all viruses can infect and impair all susceptible CD4 T cells. This leads to a diversity threshold, beyond which immunity collapses and AIDS develops. In more general terms, evolution of several viral characteristics (such as the replication rate, cell tropism, etc) can contribute to HIV disease progression.

1.6 Immune Response Dynamics: Structure of the Book

The above section outlined briefly some interesting insights that were obtained by mathematical models of HIV infection. These models focused mostly on the viral aspects of the interactions between viruses and the immune system (such as the viral turnover rate, the evolutionary dynamics of drug resistance, and immune escape). Many more interesting mathematical studies have been performed in this context, and this cannot be covered in detail here. A good summary of this work is given in [Nowak and May (2000)] and [Perelson (2002)]. In addition to the dynamics of viral infections, however, it is also important to study details about the dynamics of immune responses. What are the principles that underlie the functioning of immune responses? How do immune responses successfully clear infections, and when do they fail? How can weak immune responses be boosted by treatment? These types of questions are the focus of the book. In particular, the book explores a specific branch of the immune system: the cytotoxic T lymphocytes (CTL), which are essential in the fight against many viral infections. Concentrating on this branch of immunity allows a focused and detailed review of how mathematical models, in conjunction with experimental work, can provide insights into the biology of immune responses. The book is structured as follows. Chapter 2 reviews various mathematical models that have been used to describe CTL dynamics. We start with the simplest models that have been used earlier on, and subsequently introduce more complexity and biological realism. The models that are introduced in Chapter 2 form the basis for the rest of the book. Subsequent chapters refer to the equations set out in Chapter 2. They review important insights that have been obtained by such models. First, we explore basic principles of CTL responses such as their ability to control infections, conditions when viruses can establish persistent infections, and the concepts of immunological memory, immunodominance, and CD4 T cell help. Further chapters concentrate on the effector function of CTL. That is, different mechanisms by which CTL can fight viruses (e.g. killing infected cells versus silencing of the viral genome in infected cells), the occurrence of CTL-induced pathology, and the role of effector molecules for CTL homeostasis. These general mathematical considerations are then applied to the specific case of human immunosuppressive infections, such as HIV. We discuss how virus-induced immune impairment influences the infection dynamics, and explore how drug therapy can be used to overturn immune impairment and to induce long-term immunological control of the infection. The book concludes by looking at immune response dynamics on an evolutionary and epidemiological level. In particular, we explore, how the properties of immunological memory can affect the coevolution of parasites and their hosts.

2

Models of CTL Responses and Correlates of Virus Control

This chapter will introduce basic mathematical models that have been used to study the dynamics between virus infections and CTL responses. These models range from very simple and phenomenological to more complicated models that try to capture many biological details explicitly. This range also represents the development of these models over time, as more and more biological information became available from experimental research. Yet, even the models that are based on latest biological information still include uncertain assumptions because the appropriate biological and molecular details have not been fully worked out so far.

Nevertheless, all these models are useful. The aim of a mathematical model is not to include every molecular detail that is involved in the interactions between CTL and viruses. In fact, this amount of complexity in a mathematical model would make it very difficult to achieve any meaningful insights by analysis. Instead, the aim of the model is to capture certain biological assumptions that are thought to be key factors driving the dynamics between CTL and viral infections, and to follow them to their precise logical conclusions. This can allow us to obtain an understanding that would otherwise not be possible, to interpret experimental data, to generate new hypotheses, and to design new experiments.

The simplest and most basic question in model analysis concerns the correlates of CTL-mediated virus control. Which factors determine virus load and the degree of control? Under which conditions are infections cleared by CTL, and when are persistent infections established? We will go through a variety of mathematical models and provide a basic analysis that centers around these questions. This analysis will provide the basis for subsequent chapters that use these models in order to examine particular aspects of the interactions between CTL responses and viral infections. We start with a discussion of equations that describe the process of virus replication *in vivo*, without the presence of any immune responses. The equations describing CTL dynamics will be built on top of these basic equations of virus dynamics.

2.1 Virus Dynamics

The basic principles that underly models of virus dynamics are as follows (Fig 2.1). When viruses meet susceptible cells, they infect them. Infected cells produce new virus particles that leave the cell and find further susceptible target cells. Repeated rounds of infection result in the growth of the virus population. Growth is, however, not without bound, but is limited by the availability of target cells. Once the virus has infected most cells of the tissue, the population cannot grow further. This process can be modeled either in a very simplistic way, or the viral life cycle can be described in more detail. Both models will be described in turn.

The simplest way to model virus replication is by a density dependent logistic growth equation. This is a very common equation in ecology that is used to model the growth of populations [Lotka (1956)]. Denoting the virus population by the variable v, the model is given by the following ordinary differential equation.

$$\dot{v} = rv\left(1 - \frac{v}{\omega}\right) - av. \tag{2.1}$$

The rate of virus growth is given by the parameter r. Growth is density dependent and saturates at a "'carrying capacity"' ω. This represents target cell limitation. Finally, the virus population decays with a rate a. Therefore, this model does not distinguish between uninfected cells, infected cells, and free viruses. Instead, it captures the virus population in a single variable. Such a phenomenological model is often useful, especially if basic virus growth equations need to be incorporated into complex models. This model has two equilibria. If $r < a$, the virus fails to establish an infection and this is described by $v^* = 0$. On the other hand, if $r > a$, then an infection is established successfully and this is described by $v^* = \omega(r - a)/r$.

Now, we consider a more detailed model that distinguishes between susceptible uninfected cells x, infected cells y, and free virus v [Anderson and May (1991); De Boer and Perelson (1998); Nowak and May (2000)]. The model is given by the following set of ordinary differential equations.

$$\dot{x} = \lambda - dx - \beta xv, \tag{2.2}$$
$$\dot{y} = \beta xv - ay, \tag{2.3}$$
$$\dot{v} = ky - uv. \tag{2.4}$$

It is explained schematically in Fig 2.1. Uninfected cells are produced with a rate λ, and die with a rate d. These are basic tissue dynamics in the absence of an infection. A balance between production and death will maintain the tissue at a given size. When these susceptible cells meet free virus particles, they become infected with a rate β. The infected cells die with a rate

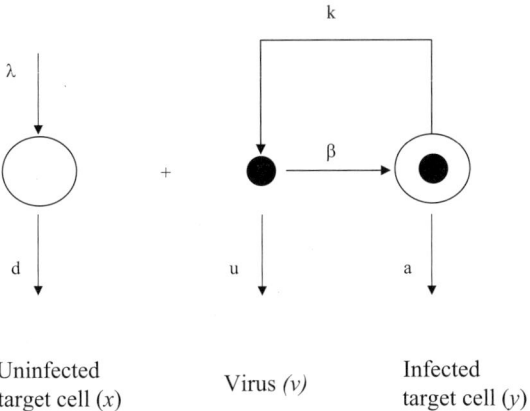

Fig. 2.1. Schematic diagram of the mathematical model (2.2–2.4). Uninfected cells are produced with a rate λ and die with a rate d. Upon encounter with virus, infection occurs with a rate β. Infected cells die with a rate a and produce free virus particles with a rate k. Free virus decays with a rate u.

a. This death rate will often be larger than the death rate of uninfected cells because viruses cause cell damage and cell death (cytopathicity or cytotoxicity). Infected cells produce new virus particles with a rate k, and the free virus particles that have been released from the cells decay with a rate u. We observe the same types of equilibria as in the simple model (2.1). Failure to establish an infection is described by $x^* = \lambda/d$, $y^* = 0$, $v^* = 0$. Successful establishment of infection is described by

$$x^* = \frac{au}{\beta k},$$
$$y^* = \frac{\lambda\beta k - dau}{a\beta k},$$
$$v^* = \frac{\lambda\beta k - dau}{a\beta u}.$$

Which outcome is achieved is determined by the *basic reproductive ratio* of the virus R_0 [Anderson and May (1991)]. This is the average number of newly infected cells produced by a single infected cell at the beginning of the infection (when almost all cells are still uninfected). If $R_0 > 1$, then one cell on average gives rise to more than one newly infected cell, and the infection can spread. If $R_0 < 1$, then one cell on average gives rise to less than one newly infected cell, and the virus population fails to spread and goes extinct. The basic reproductive ratio of the virus is given by

$$R_0 = \frac{\lambda\beta k}{dau}.$$

It is therefore determined not only by viral, but also by host parameters. For example, if there are more susceptible target cells around (higher value of λ/d), the virus can spread faster and this elevates its value of R_0. This concept of the basic reproductive ratio of the virus is described schematically in Fig 2.2, and is a very important measure that can be applied to experimental data. For example, it can tell us by how much the efficacy of certain drugs has to be increased in order to eradicate an infection from a host [Little et al. (1999); Nowak et al. (1997)], and it plays an important role in the analysis of drug resistance [Bonhoeffer et al. (1997)].

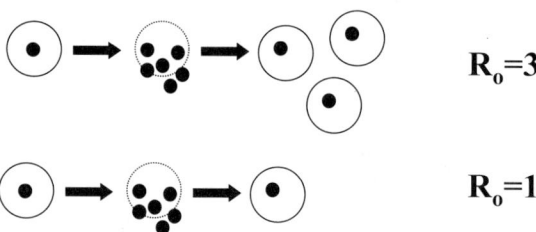

Fig. 2.2. The basic reproductive ratio of the virus R_0. This measure expresses the average number of newly infected cells produced by a single infected cell at the beginning of the infection. If $R0 > 1$, the infection becomes established. If $R0 < 1$, the infection goes extinct. If $R_0 = 1$, then one infected cells on average gives rise to one newly infected cell. This case is a border case, and irrelevant for practical purposes.

It should be noted that this model can be simplified. Free virus populations turn over at much faster rates than the population of infected and uninfected cells. Therefore, we can assume that the virus population is in a quasi steady state ($v = ky/u$), and we can rewrite model (2.2–2.4) as follows.

$$\dot{x} = \lambda - dx - \beta' xy, \tag{2.5}$$
$$\dot{y} = \beta' xy - ay, \tag{2.6}$$

where $\beta' = \beta k/u$. The basic reproductive ratio can thus be written as $R_0 = \beta' k/da$. Often this simplified version of the model is given, and the composite parameter β' is written without the prime for simplicity.

2.2 Simplest Model for CTL Dynamics

Here, we introduce the simplest way in which the dynamics of CTL responses can be modeled [De Boer and Perelson (1998); Nowak and Bangham (1996)].

2.2 Simplest Model for CTL Dynamics

We consider a single population of CTL that fights the infection and denote it by z. The essence of CTL responses is that CTL proliferate when they are stimulated by viral antigen. The expanding CTL population then fights the virus population, for example by killing infected cells. This is very similar to the dynamics between predators and prey in ecology. The CTL are the predators that grow on and kill their prey (virus). Building on the basic virus dynamics equations (2.5–2.6), this is described by the following set of ordinary differential equations.

$$\dot{x} = \lambda - dx - \beta' xy, \tag{2.7}$$
$$\dot{y} = \beta' xy - ay - pyz, \tag{2.8}$$
$$\dot{z} = cyz - bz. \tag{2.9}$$

The CTL proliferate in response to antigenic stimulation with a rate c, and die in the absence of antigenic stimulation with a rate b. The parameter c has also been referred to as the *CTL responsiveness*. CTL kill infected cells with a rate p. It is important to note that CTL can also have nonlytic activity. For simplicity we do not include this here. Instead, we will introduce such a model and analyze it in detail in Chapter 8. Further note, that naive CTL are produced by the thymus, but this is not included in the model. The reason is that the production rate of the CTL is very low, and naive CTL specific for a given virus only exist at very low numbers. Therefore, this input term can be ignored.

Assume that the basic reproductive ratio of the virus $R_0 > 1$, such that the virus can successfully establish an infection. Now we observe two possible outcomes. If $c(\lambda/a - d/\beta) < b$, then the CTL response fails to become established. This is because the CTL responsiveness c is too low to ensure sustained CTL expansion. This outcome is thus described by the following equilibrium expressions:

$$x^* = \frac{[}{K} omatsuetal.]\beta, y^* = \frac{\lambda}{[} Komatsuetal.] - \frac{d}{\beta}, z^* = 0.$$

On the other hand, if $c(\lambda/a - d/\beta) > b$, then a sustained CTL response develops, and the system converges to the following equilibrium.

$$x^* = \frac{\lambda c}{dc + \beta b},$$
$$y^* = \frac{b}{c},$$
$$z^* = \frac{c(\beta\lambda - ad) - ab\beta}{p(dc + b\beta)}.$$

In a typical simulation of this system (Fig 2.3), virus first grows and stimulates the CTL. The CTL population expands and fights the virus population. Damped oscillations occur and the system approaches its steady state. Virus load at the steady state is determined by two parameters. The rate of CTL proliferation, or CTL responsiveness c, and the death rate of CTL in the absence of antigenic stimulation b. The higher the CTL responsiveness (higher value of c), and the longer the life span of CTL in the absence of antigen (lower death rate of CTL, i.e. lower value of b), the lower virus load. These are thought to be properties of a CTL memory response, and this concept is developed further in Chapter 3. Because the model is deterministic, virus load can never be reduced to zero. Instead, however, it can be reduced to infinitely low values, which correspond to virus extinction in practical terms (average virus load is reduced to less than one virus particle).

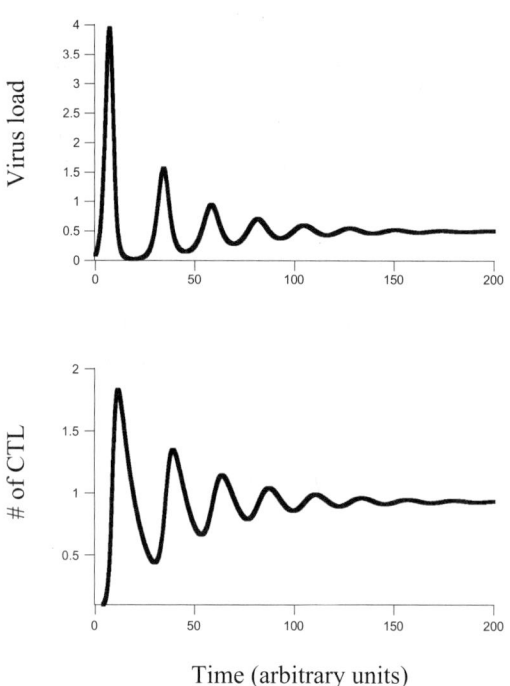

Fig. 2.3. Simulation of the simplest model that describes CTL dynamics (2.7–2.9). Virus growth is followed by CTL expansion, and CTL-mediated activity reduces virus load. Subsequent damped oscillations bring the system towards an equilibrium. The level of virus load at this equilibrium shows how well the infection is controlled. If virus load lies below a threshold, this indicates virus clearance in practical terms. After expansion, the population of CTL remains at an elevated memory level. Parameters were chosen as follows: $\lambda = 1$, $d = 0.1$, $\beta = 0.1$, $a = 0.2$, $p = 0.5$, $c = 0.2$, $b = 0.1$.

The major advantage of this model is its analytic simplicity. However, it does have some unrealistic features. The CTL proliferation term is very strong. That is, the rate of CTL expansion is directly proportional to the number of CTL, and there is no saturation. In this case, the number of CTL is only regulated by the amount of antiviral activity exerted by the CTL. Assume that the CTL have lost their ability to kill and to perform any other antiviral activity. The model predicts that the number of CTL would increase towards infinity. While a reduction in CTL-mediated antiviral activity does lead to increased numbers of CTL *in vivo* [Stepp et al. (2000)] (see Chapter 10), an increase to infinity is certainly unrealistic. As a consequence of this strong proliferation term, virus load is also only determined by immune parameters; viral parameters, such as the viral replication rate, have no influence at all. Again, this is an unrealistic feature of the model. Therefore, several variations of this model have been used [De Boer and Perelson (1995); De Boer and Perelson (1998); Wodarz and Nowak (2000b)] that aim to make this CTL expansion term more realistic. In the following sections, models that describe saturated CTL expansion will be discussed, because these will be used in later chapters. In general, however, it is important to realize that this simplified model can be used to describe CTL dynamics and to gain some meaningful insights, as long as one is aware of its limitations, and the results obtained do not depend on the unrealistic features of the equation.

2.3 Saturated CTL Expansion

The simple model of CTL dynamics (2.7–2.9) can be modified to assume that the rate of CTL expansion saturates as the number of CTL grows to relatively high numbers. This is expressed by the following differential equation [Wodarz and Nowak (2000b)].

$$\dot{z} = \frac{cyz}{\epsilon z + 1} - bz. \tag{2.10}$$

The level at which CTL expansion saturates is expressed in the variable ϵ. The condition for the establishment of the CTL response is the same as in the previous model (2.7–2.9). If a sustained CTL response becomes established, the system converges to the following equilibrium.

$$x^* = \frac{\beta b(\epsilon a - p) - pcd + \sqrt{[\beta b(\epsilon a - p) - pcd]^2 + 4\beta^2 b\epsilon\lambda cp}}{2\beta^2 b\epsilon},$$

$$y^* = \frac{\lambda - dx^*}{\beta x^*},$$

$$z^* = \frac{\beta x^* - a}{p}.$$

This model has similar properties compared to the simple model discussed above (2.7–2.9). A higher CTL responsiveness c and a longer life span of the

CTL in the absence of antigen (lower value of b), correlate with lower virus load. In addition, however, virus load is also a function of viral parameters, most importantly the replication rate of the virus β. An increase in the parameter β leads to an increase in viral load up to an asymptote. Also, if the CTL do not have any antiviral activity ($p = 0$), then the number of CTL does not increase to infinity, but only up to a defined value. Thus, inclusion of the saturation term eliminates the biologically unrealistic features of the simpler model (2.7–2.9). However, this comes at the cost of having more complicated equilibrium expressions that make the model less tractable by analytical means.

If saturation already occurs at lower numbers of CTL (high value of ϵ), it is possible to use a simpler version of this model, given by [Wodarz et al. (2002)].

$$\dot{z} = cy - bz \quad (2.11)$$

In this model the rate of CTL expansion is simply proportional to the amount of antigen, but not to the number of CTL. The rate of CTL expansion is therefore weakened. In this model, the CTL response can never go extinct. Instead, if the CTL responsiveness c is low, the CTL persist at low levels. Therefore, if $R_0 > 1$, there is only one stable equilibrium, and this is given by the following expressions.

$$x^* = \frac{\beta ba - pcd + \sqrt{[\beta ba - pcd]^2 + 4\beta^2 b \lambda cp}}{2\beta^2 b},$$

$$y^* = \frac{\lambda - dx^*}{\beta x^*},$$

$$z^* = \frac{\beta x^* - a}{p}.$$

These equilibrium expressions have qualitatively the same properties as those of the saturation model discussed above (2.10). There are several other variations to describe CTL dynamics, and they all lead to more realistic features compared to the simplest model (2.7–2.9). Since they will not be used in this book, however, they will not be discussed here. The interested reader is referred to [De Boer and Perelson (1998)]. A note of caution: while these models do exhibit more realistic behaviors, the saturation terms are arbitrary and are not based on any specific biological detail. This has to be kept in mind when interpreting modeling results.

2.4 Precursor Versus Effector CTL

The models discussed so far capture the CTL in a single population z. In reality, however, the population of CTL can be divided into at least two subpopulations: CTL precursors or CTLp and CTL effectors or CTLe (Fig 2.4).

2.4 Precursor Versus Effector CTL

CTLp do not have any antiviral activity, while CTLe do have antiviral activity. Naive CTL (which have never seen antigen before), exist as precursors. When they become stimulated by antigen, the population of CTLp expands. This culminates in the differentiation into CTL effectors that fight the virus. Memory CTL are again precursor CTL without antiviral activity. In order to attain antiviral activity, the memory CTL need to be stimulated again. Mathematical models have been constructed that take into account this subdivision of the CTL [Wodarz et al. (1998); Wodarz et al. (2000b); Wodarz et al. (2000c)]. Denoting the population of CTL precursors by w and the population of CTL effectors by z, the model is given by the following set of differential equations.

$$\dot{w} = cyw(1-q) - bw, \qquad (2.12)$$
$$\dot{z} = cqyw - hz. \qquad (2.13)$$

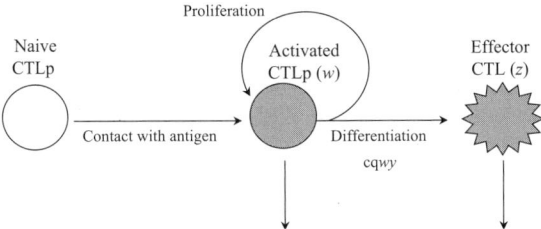

Fig. 2.4. Schematic representation of the model (2.12–2.13), which distinguishes between two subpopulations of CTL: the precursors, CTLp, and the effectors, CTLe. Precursors become activated by antigen, proliferate, and differentiate into effectors. CTLp do not have any antiviral activity, but CTLe do.

Upon contact with antigen, CTLp proliferate at a rate cyw and differentiate into effector cells at a rate $cqyw$. CTL precursors die at a rate bw, and effectors die at a rate hz. In this model, CTL memory lies in the population of precursors w. If the life span of the CTLp is long in the absence of antigen (low value of b), then the CTLp population persists at elevated levels for prolonged periods of time after acute infection, as observed *in vivo*. On the other hand, the life span of the effector CTL is assumed to be short (relatively high value of h), because prolonged effector activity can have damaging consequences for the host.

Experimental data indicate that upon expansion, CTL first differentiate into effectors before assuming the memory phenotype again [Kaech et al. (2002)]. This is not explicitly captured in the equations in order to preserve analytical simplicity. Rather, this model is phenomenological. The CTL expansion and differentiation process that gives rise to effectors and eventually increases the number of memory CTLp is captured in the term cyw (resting memory CTL → activated proliferating CTL → effector CTL → resting memory CTL). We assume that effector CTL are simply proportional to the number of memory CTL and the amount of antigen (expressed by the term $cqyw$). That is, we make a quasi steady state assumption.

Assume that the the basic reproductive ratio of the virus is greater than one. A sustained immune response can become established if $c(1-q)(\lambda/a - d/\beta) > b$. In this case, the system converges to the following equilibrium.

$$x^* = \frac{\lambda c(1-q)}{dc(1-q) + b\beta}, \tag{2.14}$$

$$y^* = \frac{b}{c(1-q)}, \tag{2.15}$$

$$w^* = \frac{z^* h(1-q)}{bq}, \tag{2.16}$$

$$z^* = \frac{\beta x^* - a}{p}. \tag{2.17}$$

According to these expressions, virus load is reduced by a high responsiveness and a long life span of the *memory* CTLp. Therefore, this model indicates that the memory phenotype of the CTL is crucial for virus control. The level of virus load is independent of the parameters of the effector CTL. These notions are further developed in Chapter 3.

Note that the approach to the equilibrium can be more complex compared to the simpler models discussed earlier in this chapter. For low values of b (long life span of CTL in the absence of antigen), the system takes a long time to equilibrate. After an initial transient phase, the dynamics lead to a quasiequilibrium $(y^\wedge)$ at which virus load only decays at a very small rate. Virus load at the quasiequilibrium is higher than at the true equilibrium, but has similar properties. Hence, virus load at the quasiequilibrium can be approximated by $y^\wedge = \alpha y^*$, where $\alpha > 1$. After a time period of $1/b$, the system approaches the true equilibrium y^*. This can have implications for the control of persistent infections at low levels (see Chapter 3).

2.5 Programmed CTL Proliferation

The models discussed so far assumed that CTL require continuous antigenic stimulation for cell division and proliferation. That is, if antigenic stimulation was withdrawn, the CTL would stop to proliferate. However, this notion turned out not to be true. Instead, a single encounter with antigen triggers a program of CTL expansion and differentiation that is independent of further antigenic stimulation events [Kaech and Ahmed (2001); Mercado et al. (2000); van Stipdonk et al. (2003); van Stipdonk et al. (2001)] (Fig 2.5).

This is referred to as programmed proliferation. It is thought that the CTL undergo approximately 7-10 cell divisions that result in the generation of effector cells, and subsequently in the differentiation into memory cells. Even if antigen is withdrawn at any time during this process, the proliferation and differentiation program is still completed. If this process does not result in the clearance of the pathogen, the memory CTL are reactivated and further expansion occurs. It has been argued that the existence of the program significantly alters the properties of CTL dynamics, and that conclusions reached by earlier models are not valid anymore. Hence, it is important to construct a model of programmed CTL proliferation and to compare its properties to those of the continuous stimulation models discussed above. This is done as follows.

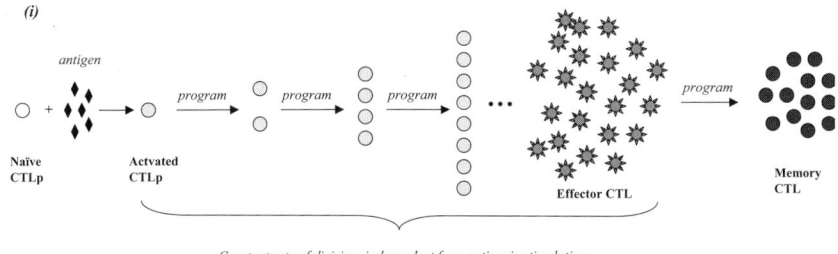

Fig. 2.5. Schematic representations of (i) the programmed proliferation and (ii) the continuous stimulation concepts.

The mathematical model that describes programmed CTL proliferation [Wodarz and Thomsen (2005)] contains the following variables: resting and memory CTL m; newly activated CTL m_0; activated CTL that have undergone i ($i = 1..n$) cell divisions m_i; and effector CTL z. It is given by the following set of differential equations.

$$\dot{z} = 2rm_{n-1} - \gamma z - \delta z, \tag{2.18}$$

$$\dot{m}_{n-1} = 2rm_{n-2} - rm_{n-1}, \tag{2.19}$$

$$\ldots \quad (2.20)$$
$$\dot{m}_1 = 2rm_0 - rm_1, \quad (2.21)$$
$$\dot{m}_0 = \alpha y m - rm_0, \quad (2.22)$$
$$\dot{m} = \gamma z - \alpha y m - \epsilon m. \quad (2.23)$$

We start with a population of resting CTL, denoted by m. Upon antigenic encounter, these cells become activated at a rate $\alpha y m$. These activated cells are denoted by m_0. Following activation, the CTL undergo n rounds of proliferation, and this is independent of antigenic stimulation. Proliferation occurs with a rate $2rm_i$, where m_i denotes CTL that have undergone i divisions ($i = 1..n-1$). The nth division gives rise to effector cells, denoted by z. They can kill infected cells (or alternatively have nonlytic activity). Effectors die at a rate δz and differentiate into memory cells at a rate γz. Memory cells are again denoted by m, since they are resting. If the virus is not cleared after this first round of programmed proliferation, the memory cells are reactivated according to the same principles as described above and undergo another round of programmed proliferation. The only difference is that memory cells are characterized by an elevated activation and proliferation rate compared to naive cells (higher values of α and r, respectively). The model is easily adapted to assume that effectors are generated after n_E cell divisions ($n_E < n$), and that effectors subsequently undergo further divisions until they have divided n times (9).

The acute infection dynamics are obviously different in the program model compared to those observed in models that assume continuous antigenic stimulation. This is because during this phase, the response is on autopilot, and is not influenced by antigen. The acute phase of infection can be defined by the first round of programmed CTL proliferation that ends with the generation of memory cells (that is, memory cells are not restimulated; more details in Chapter 4). Here we concentrate on the long-term dynamics. In this case, there is persistent infection and restimulation of memory cells. That is, the virus is not cleared after the first round of programmed proliferation. The properties of the program model will be compared to those of the models that assume continuous antigenic stimulation in this context.

Recurring rounds of CTL proliferation will be induced, and the system will eventually converge towards a steady state. Virus load at this steady state determines the degree of virus control. The equilibrium expressions are given as follows.

$$x^* = \frac{a + gy^*m}{\beta},$$
$$y^* = \frac{\epsilon}{\left(\frac{2^n \gamma}{\gamma + \delta} - 1\right)\alpha},$$

$$m^* = \frac{1}{gy^*}\left(\frac{\lambda}{d/\beta + y^*} - a\right),$$

$$z^* = \frac{gy^* m^*}{p},$$

$$m_0^* = \frac{\alpha y^* m^*}{r},$$

$$m_i^* = 2^i m_0^*, 1 < i < n-1,$$

where $g = (2^n p\alpha)/(\gamma + \delta)$.

Virus load is determined by a number of immunological factors. A high activation rate of memory CTL (high value of α) and a long life span of memory CTL in the absence of antigen (low value of ϵ) contribute to low virus loads. Also, the higher the number of CTL proliferations (higher value of n), the lower virus load. The number of CTL at the steady state is mainly determined by the rate of antiviral activity p. The lower the rate of CTL-mediated antiviral activity, the higher the number of CTL. Interestingly, these properties are almost identical to the properties derived from mathematical models that assume that CTL proliferation requires continuous antigenic stimulation. In fact, the continuous stimulation models are a special case of the programmed proliferation model in which the program is executed with a very fast rate. This is shown mathematically as follows.

Assume that upon antigenic stimulation, the program is executed at a very fast rate (high values of r), and that the turnover of activated and effector CTL is significantly faster than the turnover of memory cells. In this case, the programmed proliferation model can be reduced to a single equation for the memory CTL. It is given by

$$\dot{m} = \left(\frac{\gamma 2^n}{\gamma + \delta} - 1\right)\alpha y m - \epsilon m. \qquad (2.24)$$

This is basically the same equation as the simplest continuous stimulation model (2.7–2.9), where the CTL responsiveness is given by

$$c = \left(\frac{\gamma 2^n}{\gamma + \delta} - 1\right)\alpha.$$

Therefore, the single CTL population in the continuous stimulation model should be considered as the population of memory CTL. The effector CTL population can be assumed to be in quasi steady state and is given by $z = 2^n \alpha y m/(\gamma + \delta)$. Consequently, the rate of killing is described by $p'y^2 m$, where $p' = 2^n \alpha/(\gamma + d)$. The killing term is proportional to the square of virus load (y) because the generation of effector cells from memory cells is proportional to virus load. In the simple continuous stimulation models (e.g. (2.7–2.9)), the killing term is only linearly proportional to virus load because the model does not distinguish between memory and effector CTL. More complicated continuous stimulation models that distinguish between memory and effector CTL (2.12–2.13) have the killing term essentially proportional to the square of virus load.

Given the similarities in the steady state properties of the programmed proliferation and the continuous stimulation models, we ask the question why programmed proliferation exists. The answer is that the equilibrium outcome of the model does not tell the whole story (Fig 2.6). Instead, the dynamical approach to the equilibrium provides interesting insights. We compare the properties of programmed proliferation (2.18–2.23) to those of continuous antigenic stimulation (2.24). We distinguish between two scenarios. First we assume a strong CTL memory response that gives rise to a very low virus load at equilibrium (clearance). Then we assume a weaker CTL response that correlates with the persistence of higher virus load at equilibrium.

Assume that the CTL response is strong (Fig 2.6i). Consider the continuous stimulation model first. Initially, the CTL response can be very efficient at stopping viral growth and reducing virus load. This is because higher virus load increases the rate of CTL proliferation. As virus load declines, however, the effectiveness of the response becomes greatly diminished. This is because generation of effectors requires constant antigenic stimulation, and the amount of antigen is low. Consequently, the dynamics enter a phase where virus load settles at a level that is significantly higher than the predicted equilibrium and where virus load declines at a very slow rate (Fig 2.6i, quasiequilibrium discussed above). Now, consider the programmed proliferation model. In this case, CTL divisions are independent of antigenic stimulation. This provides an initial disadvantage: as virus load grows, the increased level of antigenic stimulation does not result in faster CTL expansion and the virus can more easily grow to high levels and cause acute pathology. As virus load is reduced to low levels by the CTL, however, the CTL can keep dividing despite the small amounts of antigenic stimulation. Thus, in contrast to the continuous stimulation model, production of effectors does not slow down abruptly as virus load drops. Consequently, CTL-mediated pressure is maintained at low virus loads and this results in efficient reduction of the virus population to very low levels or extinction. Thus, clearance can occur before the system converges to an equilibrium (Fig 2.6i). According to these arguments, we observe a tradeoff between the ability of the CTL to clear an infection and the ability of CTL to reduce acute phase symptoms. If the CTL are more efficient at virus clearance, they are less efficient at containing acute virus load, and vice versa. Thus, to optimize the fitness of the host, there should be enough programmed divisions to ensure clearance, but no more such that acute pathology is limited. We hypothesize that the 7-10 antigen independent CTL divisions observed in experimental data represent this optimum.

Now assume a weaker CTL memory response (Fig 2.6ii). In this case, equilibrium virus load is higher, which can correspond to persistent infection. The same equilibrium is reached, both in the continuous stimulation and the

2.5 Programmed CTL Proliferation

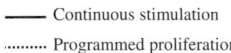

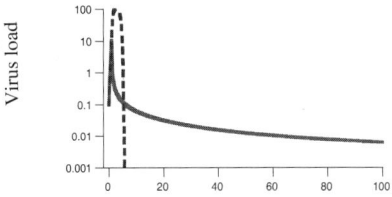

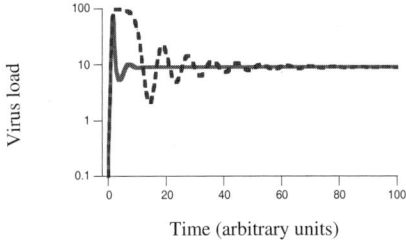

Time (arbitrary units)

Fig. 2.6. Comparison between the programmed proliferation model (2.18–2.19) and a continuous stimulation model (2.24). We distinguish between virus dynamics in the presence of a strong CTL response (that can lead to low virus loads and clearance), and a weak CTL response that results in persistent infection. (i) If the response is strong, programmed proliferation leads to higher peak virus loads during acute infection, but leads to more efficient clearance compared to the continuous stimulation scenario. In the continuous stimulation model, virus load only converges very slowly to its equilibrium value, and this hinders clearance. The reason is that effector production relies on continuous encounter with antigen, which is limiting at low loads. (ii) If the CTL are weak and persistent infection is established, both the programmed proliferation model and the continuous stimulation model have similar properties: they converge to the same equilibrium. The program model takes slightly longer to converge to the equilibrium because there is delay between induction of the response and the generation of effectors. Parameters were chosen as follows: (i) $\lambda = 10$, $d = 0.1$, $\beta = 0.05$, $a = 0.1$, $r = 5$, $\gamma = 1$, $\alpha = 0.01$, $\delta = 0.5$, $p = 0.2$, $\epsilon = 0.001$; (ii) $\lambda = 10$, $d = 0.1$, $\beta = 0.05$, $a = 0.1$, $r = 5$, $\gamma = 0.1$, $\alpha = 0.005$, $\delta = 1$, $p = 0.1$, $\epsilon = 1$.

programmed proliferation model. The outcome of the dynamics does not depend significantly on these model differences. Thus, if the CTL fail to resolve the infection, the continuous stimulation and the programmed proliferation models give rise to similar predictions.

In summary, the process of programmed CTL proliferation can enhance the ability of the response to clear viral infections because it allows elevated CTL effector activity to persist at low virus loads. In the context of persistent infections, however, the properties of programmed proliferation and continuous stimulation are very similar. Therefore, it is likely that results obtained from continuous stimulation models regarding CTL responses against persistent infections remain robust in the context of programmed proliferation. Many of the results that are based on continuous stimulation models are therefore valid.

2.6 Summary

This chapter has summarized a variety of mathematical models that have been used to describe the development of CTL responses against viral infections. We started with the simplest model, and subsequently added complexity to account for more biological details. We discussed basic properties regarding virus control in the context of these models, and this sets the stage for specific aspects of CTL dynamics that will be discussed throughout the book. It is important to realize that all of the models contain assumptions that represent simplifications, or that have not yet been possible to verify by experiments. Therefore, when interpreting model predictions, it is essential that results do not depend on uncertain mathematical expressions used in the equations.

3
CTL Memory

Immunological memory is a central characteristic of the immune system [Ahmed and Gray (1996); Doherty et al. (1996); Zinkernagel (1996); Zinkernagel (2000a)]. Nevertheless, there is no simple definition of this concept. The easiest definition is on the functional level: a host is protected more efficiently against a second infection if it has previously been infected with the same pathogen, and survived the infection. Before the host has encountered a pathogen it is said to be *naive*. This means that the number of immune cells that are specific for this pathogen is relatively low. When the host is infected with this pathogen for the first time (*primary infection*), the population of specific immune cells expands, fights the pathogen, and subsequently settles around a relatively stable level that is much higher than in the naive host. This population of cells is referred to as *memory cells*, and memory persists in the long-term after pathogen clearance. Upon reinfection with the same pathogen (*secondary infection or rechallenge*), this population of memory cells can react more efficiently against the invading virus compared to a naive host. Consequently, the host suffers less or no pathology. Memory is found in all adaptive branches of the immune system: B cells (or antibody) responses, as well as T cell responses. Vaccinations rely on the generation of memory: the immune system is artificially exposed to a pathogen (or an immunogenic part of the pathogen), and this results in the generation of immunological memory and in protection against infection. While the population of memory cells is maintained in the long-term after clearance of a pathogen, it declines slowly, and this can lead to a loss of protection over time. It is unclear for how long hosts are protected against reinfection, and this may vary from case to case.

Apart from these general considerations, uncertainties remain regarding the nature of immunological memory, and how protection is achieved [Bruno et al. (1995); Zinkernagel (2002b)]. On the one hand, the elevated level of immune cells that remains after resolution of an infection can be directly responsible for the enhanced protection: the presence of a larger army of cells is more efficient at fighting a pathogen. On the other hand, memory cells have special traits. They are characterized by specific surface markers, and have an enhanced ability to become activated and to proliferate in response to antigenic stimulation. This also adds to their protective capacity. The mechanisms by which protective memory cells are maintained in the long-term are debated. Some immunologists have argued that protection requires that memory cells be continuously exposed to small amounts of antigen that might remain in the system even after resolution or clearance of a pathogen [Beverley (1990); Gray and Matzinger (1991); Kundig et al. (1996a); Kundig et al. (1996b)]. On the other hand, it has been shown that memory cells can survive for a long period of time in the absence of any exposure to antigen [Ahmed and Gray (1996); Jamieson and Ahmed (1989); Lau et al. (1994); Murali-Krishna et al. (1999)]. It is possible that while memory cells are indeed long-lived in the absence of antigen, they need to be exposed to antigen periodically in order to remain in a higher state of activation and thus to be able to fight the pathogen immediately upon infection [Kundig et al. (1996a); Kundig et al. (1996b)].

There are also important differences between the the properties of memory in different branches of the adaptive immune system. The role of memory is clearest in the context of B cells and antibodies. All successful vaccines are based on antibody responses. Antibodies are passed from mother to child and provide immunological protection during the first six months of life. Memory T cells, on the other hand, are not transferred from mother to child. Among T cells, the CTL are thought to have the main protective role since they exert direct antiviral activity, while helper T cells are mostly involved in the activation of effector responses. But even with CTL, the protective role is unclear. While memory CTL can persist in the long-term without antigenic stimulation, their ability to protect varies and can depend on the route of infection [Ehl et al. (1997); Kundig et al. (1996a); Kundig et al. (1996b)]. Resting memory CTL are parked in the lymph nodes. Therefore, they are inefficient at reacting to pathogens that enter through the periphery, but are more efficient at reacting to pathogens that enter directly into the blood. Memory CTL might need periodic exposure to antigen in order to exit the lymph nodes, to patrol the entire body, and to provide immediate protection against reinfection. CTL based vaccines have not so far been successful on a larger scale.

This chapter concentrates on memory in the context of CTL responses against viral infections. It reviews how mathematical models have helped researchers to gain a better understanding of the role of CTL memory. It starts with a brief section that summarizes the current knowledge about the generation of CTL memory in viral infections. It then discusses how these processes can be captured by mathematical models and reviews the insights gained from the modeling approaches. The dynamics both during primary and secondary challenges will be examined. It will be demonstrated how insights gained from the mathematical models can be applied to interpret experimental data.

3.1 The Generation of CTL Memory: Biological Background

When a host becomes infected with a virus, a CTL response is usually triggered (Fig 3.1). The population of virus specific CTL expands to a peak after which a contraction phase is observed. Following the contraction phase, the population of CTL settles at a stable memory level that only declines at a very slow rate (Fig 3.1). How is memory generated? The differentiation pathway from naive cells to memory cells is not entirely clear (Fig 3.1). Several experimental studies have suggested that differentiation into memory cells and into effector cells occurs by different pathways and is influenced by cytokines [Manjunath et al. (2001); Sallusto and Lanzavecchia (2001)]. Recent evidence, however, points toward a linear differentiation pathway [Kaech et al. (2002); Swain (2003); Veiga-Fernandes et al. (2000); Weninger et al. (2002); Wherry et al. (2003b)] (Fig 3.1): naive cells expand and become effector cells that have antiviral activity. The effector cells may die, or they may differentiate into memory cells that are long-lived in the absence of antigen. Upon secondary encounters with antigen, memory cells become activated and can again give rise to effectors. More detailed work has distinguished between two classes of memory cells (Fig 3.1): central memory cells, and effector memory cells [Wherry et al. (2003b)]. Central memory cells express certain surface receptors that allow homing of the cells to the lymph nodes. Central memory cells seem to have the greatest capacity to persist in the absence of antigenic stimulation and to react against secondary challenges. Effector memory cells lack these homing receptors and are located primarily in the blood, spleen, and nonlymphoid tissues. It is thought that effector memory cells are an intermediate stage, and that they differentiate into central memory cells when the infection is cleared. In case of a persistent infection, it is thought that central memory cells are not generated. Instead, the system cycles between effector memory and effector CTL, as a result of the continuous antigenic drive.

3.2 Mathematical Models of CTL Memory

CTL memory can be examined with mathematical models in a variety of ways. One approach is to use a detailed and mechanistic description of the CTL differentiation pathway. Such an approach has been explored in Chapter 2 in the context of programmed CTL proliferation (2.18–2.23). This model described antigen-mediated activation of naive CTL, followed by expansion, generation of effector activity, and differentiation into memory cells. The memory cells can in turn be reactivated and repeat this process. Such a model would explicitly describe the linear differentiation pathway outlined above.

Another approach is to use a simplified phenomenological model [Wodarz et al. (2000b); Wodarz et al. (2000c)]. As described in Chapter 2, the detailed

(i)

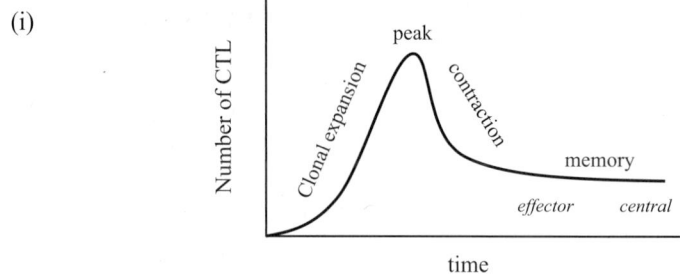

(ii)

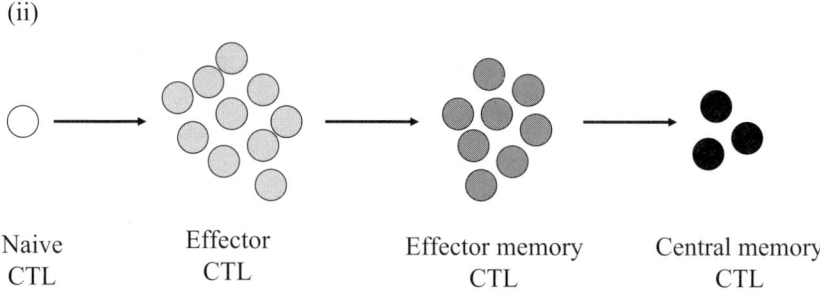

Fig. 3.1. Schematic diagram that outlines the development of CTL responses. (i) Upon infection, naive CTL become activated and clonal expansion occurs until the number of CTL has reached a peak. This leads to the differentiation into effector cells that can fight the virus population. After the peak, we observe a phase of contraction, where the number of CTL shrinks. This tends to coincide with reduction of virus load to low levels. As the infection is resolved, the number of CTL settles at a stable and elevated memory level. At first, the CTL differentiate into effector memory cells that still have a relatively high state of activation, and subsequently differentiate into resting central memory cells. (ii) The linear differentiation of CTL involves the generation of effectors, effector memory cells, and central memory cells. If antigen persists beyond this point, the memory cells can become reactivated to give rise to further effectors.

CTL differentiation model can be approximated by a simple model that describes the turnover of memory CTL in response to antigen (2.12–2.13). We explore this simplified approach in this chapter because it captures essential characteristics of memory while allowing analytical simplicity: longevity in the absence of antigenic stimulation, and a fast rate of activation. The model distinguishes between memory CTL w and effector CTL z. The long life span of memory CTL is expressed by a low value of b (slow death rate of memory CTL), while a high activation and proliferation rate of CTL is expressed by a high value of c.

3.3 Resolution of Primary Infection: Mathematical Predictions

Here we examine the conditions required to result in CTL-mediated clearance of viral infections. Because our model is deterministic, the CTL response cannot reduce virus load y to exactly zero, although virus load may be reduced to very low levels. Hence we define an extinction threshold y_{ext}. If $y < y_{\text{ext}}$, the virus population has gone extinct. Note that this threshold is arbitrary. In reality it will depend on a complex balance between host and viral parameters as well as stochastic events. In general, the lower the virus load predicted by the model, the higher the chances of viral clearance.

We start by looking at the determinants of virus load at equilibrium. According to the model, two factors can suppress virus load: (i) long-term persistence of memory CTL in the absence of antigen (low b); and (ii) a high activation rate of memory CTL (high c). Note that virus load is independent of effector parameters. Therefore, the model gives us the following important insights: Clearance or resolution of infection depends crucially on the central characteristics of CTL memory. This means that memory not only is important in the protection against secondary challenges, but may be equally important for ensuring that a primary infection is resolved. The reason is as follows (Fig 3.2). Assume that CTL are not long-lived in the absence of antigen and that they require antigenic stimulation to be maintained. In this case, the CTL response can initially drive virus load down to low levels. As virus load declines, however, the number of CTL also declines significantly because memory is weak. As the number of CTL declines, immunological pressure is lost at low virus load, and this allows the virus to grow back and to establish a persistent infection. This persistent infection will provide increased antigenic drive, and this maintains an ongoing CTL response on top of the persistent infection. The situation is different if CTL are maintained in the absence of antigen. Now CTL-mediated pressure remains strong as virus load declines to low levels, and this results in the extinction of the infection. The exact time span for which the memory CTL have to be maintained in the absence of antigen in order to control an infection is a complex function of host and viral parameters, as well as stochastic factors determining extinction events.

It is important to point out that while the outcome of infection is dramatically different in the two scenarios, the CTL dynamics can look very similar. In both cases, CTL remain in the long-term. In the case of strong memory and virus clearance, the CTL remain because they are long-lived in the absence of antigen and have the upper hand over the infection. If CTL memory is weak, the CTL persist in the long-term because they are just maintained on top of the high antigen levels.

These results have been obtained from analyzing the determinants of virus load at equilibrium. As described in Chapter 2, however, this equilibrium is ony reached after a relatively long period of time if the life span of memory CTL is long in the absence of antigen. Instead, the system settles around a

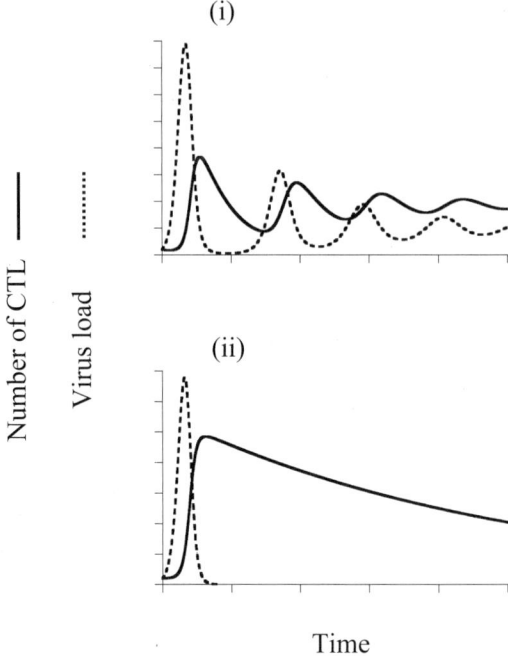

Fig. 3.2. Schematic time series that illustrates how the quality of CTL memory can decide whether an infection persists or is cleared form the host. (i) If memory is not long-lived in the absence of antigen, virus load is initially reduced to low levels during primary infection, but subsequently grows back as the CTL response diminishes at low virus loads. The result is persistent infection, where a continuous CTL response is maintained by the ongoing antigenic drive. (ii) If memory CTL are long-lived in the absence of antigen, CTL-mediated pressure is maintained at low virus loads, and this ensures that the virus is driven to extinction. In this case, the CTL persist in the long-term because they can survive for extended periods of time in the absence of antigenic stimulation. Note that a schematic time series diagram has been used to illustrate this point, and not an actual model simulation. As explained in the text, the model (2.12–2.13) is simplified in order to obtain clearer analytical results. While equilibrium analysis provides all the necessary insights, the dynamics with which the equilibrium is approached can be unrealistic due to the simplification. Since our focus is not on these dynamics, but on the long-term properties, the simplification is justified.

quasiequilibrium that is higher than the true equilibrium. At this quasiequilibrium, virus load only declines with a very slow rate. The dependence of the quasiequilibrium on the parameters of the model is, however, qualitatively the same as the dependence of the true equilibrium on model parameters (Chapter 2). Therefore, the general insights gained from the equilibrium analysis still hold. In addition, note that it is possible that virus load is cleared during the initial oscillations of the virus population, before equilibrium has been reached. Analysis of these initial oscillations again indicates that such "dynamic elimination" is enhanced by the presence of strong memory. But it is important to keep in mind that even in the absence of strong memory, dynamic elimination may occur if the infection is not very invasive, although with delayed kinetics.

3.4 Resolution of Primary Infection: Experimental Studies

The basic prediction resulting from the model is that antigen-independent persistence of memory cells may be important for CTL-mediated viral clearance during primary infection. One way to test this hypothesis is to analyze a situation in which the primary CTL response to a viral infection is intact, while memory is impaired. Such a situation is given in MHC class II- and CD40L deficient knockout mice infected with LCMV [Borrow et al. (1996); Borrow et al. (1998); Thomsen et al. (1996); Thomsen et al. (2000); Thomsen et al. (1998)]. These mutants essentially lack CD4 T cell help. While it is thought that a normal primary CTL response can be generated in the absence of CD4 T cell help, memory fails [Shedlock and Shen (2003); Sun and Bevan (2003); Sun et al. (2004)]. (The relationship between CD4 cell help and CTL responses will be explored in detail in Chapter 4). During the early phase of the infection, both wild-type and mutant animals reduce virus load to undetectable levels (Fig 3.3). However, in contrast to wild-type mice, virus load reemerges to high levels in the mutant mice about a month after infection. Resurgence of virus load is associated with lack of a significant memory CTL response in mutant animals (Fig 3.3). Interpretation of these data strongly supports the theory that CTL memory may be necessary to successfully resolve an infection. However, caution is required because the knockout mice also lack efficient B cell responses that may contribute to the overall immunological control of LCMV [Planz et al. (1997); Thomsen and Marker (1988)].

Similar results were also obtained from other mouse infections. Helper-deficient mice (that are thought to have defective memory) were infected with influenza virus [Belz et al. (2002)] (Fig 3.4). While mutant mice still managed to clear the virus eventually, it occurred with delayed kinetics. In accordance with the model, the rate of virus decline was initially similar for mutant and wild-type mice, but as virus load was reduced to low levels, the rate of virus decline slowed down significantly in mutant animals (Fig 3.4). This shows that in the absence of memory, CTL-mediated pressure may be lost at low virus loads and this may compromise the ability to resolve the infection. Another respiratory infection of mice is gamma herpes virus. It cannot be cleared from the host because it establishes latent infection in B cells. The immune system may either control it at low levels, or high virus loads are observed. CTL have been shown to contribute significantly to virus control. Helper-deficient mice

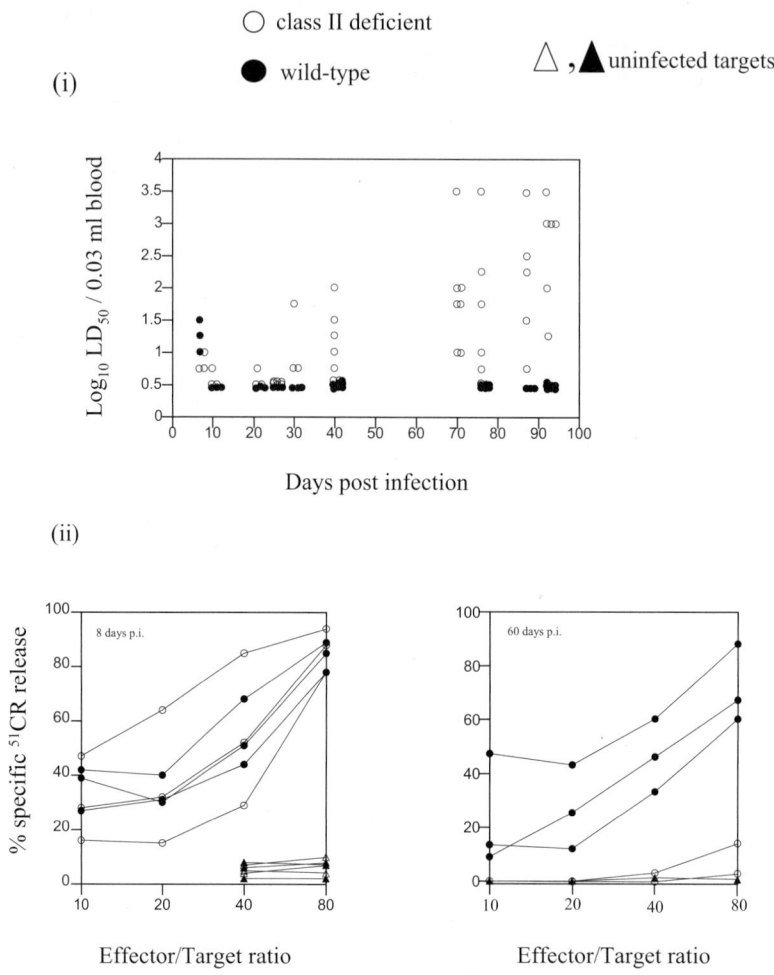

Fig. 3.3. Data replotted from Christensen et al. (1994) and Thomsen et al. (1996). (i) Dynamics of LCMV infection in class-II-deficient mice compared to wild-type mice. The y-axis is a measure of virus load. Although initial control of the virus is similar in wild-type and mutant, blood virus titers resurge in class-II-deficient mice. (ii) The primary CTL response in wild-type and mutant mice is similar. However, class-II-deficient mice show lack of a CTL memory response. The y-axis is a measure of CTL-mediated lysis. Lysis is measured at different ratios between CTL effector cells and target cells (effector:target ratio on the x-axis).

were infected with gamma herpes virus [Sarawar et al. (2001)]. Virus load was initially reduced to low numbers, but subsequently resurged. Mice were then treated with anti-CD40 antibodies. This treatment can activate antigen presenting cells that can then deliver help without the need for CD4 T

3.5 Protection Against Rechallenge: Mathematical Predictions

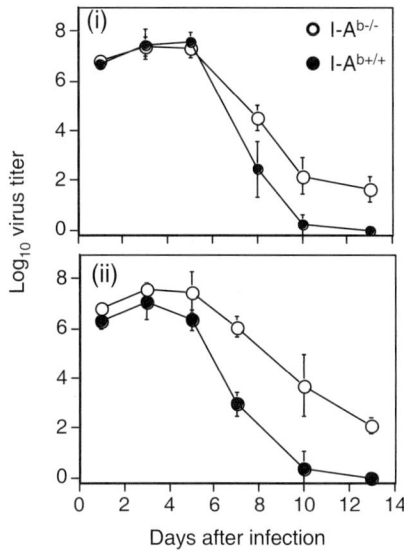

Fig. 3.4. Data replotted from Belz et al. The graphs show lung virus titers following primary (i) and secondary (ii) challenge of wild-type $(I - A^{b+/+})$ and class-II-deficient $(I - A^{b-/-})$ mice with murine influenza virus. While both mutant and wild-type mice eventually clear the infection, the kinetics of clearance are markedly slower in helper-deficient compared to wild-type mice.

cells. Therefore, this treatment made it possible to generate memory. Following treatment, virus load dropped to significantly lower levels [Sarawar et al. (2001)]. This indicates that generation of memory resulted in a reduction of virus load, as suggested by the model. Interestingly, the number of CTL was similar before and after treatment [Sarawar et al. (2001)]. Based on the model we can provide the following explanation for this observation: before treatment, CTL were short lived in the absence of antigen. Consequently, virus load was high, and CTL were maintained by antigenic drive. After treatment, memory CTL were generated that are long-lived in the absence of antigen. Consequently, virus load was reduced and the number of CTL was maintained because they could persist in the absence of antigenic stimulation.

3.5 Protection Against Rechallenge: Mathematical Predictions

We now turn to the question whether and under what circumstances CTL memory is protective against secondary infection by a virus. This is the context in which memory is traditionally considered. How is protection defined in the model? CTL memory can be considered protective if it reduces the peak virus load upon secondary challenge compared to primary challenge. This may correlate with a reduction of clinical symptoms experienced. We will consider two cases: First, we assume that the primary infection was cleared before the secondary challenge occurs. After that, we examine the situation in which the virus persists at low (and possibly undetectable) levels at the time when the secondary challenge occurs.

3.5.1 Protection After Virus Elimination

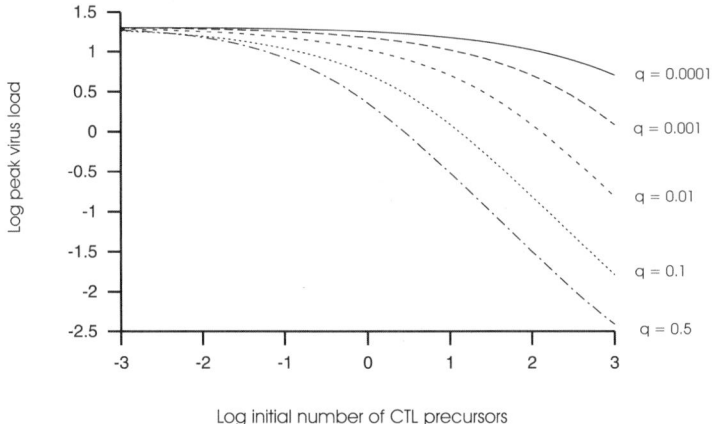

Fig. 3.5. Protection against secondary challenge after virus clearance. Plots are derived from model (2.12–2.13). Peak virus load upon secondary challenge decreases with an increase in the initial number of memory CTL precursors. However, this decrease in peak virus load is only significant if the rate of effector cell production is fast and the delay between virus entry and effector cell production is minimal (expressed in the value of q). Baseline parameters were chosen as follows: $\lambda = 10$, $d = 0.1$, $\beta = 0.001$, $a = 0.5$, $p = 1$, $c = 0.1$, $b = 0.001$, $h = 0.1$.

We assume that memory CTL are present at elevated levels following clearance of a primary infection. We further assume that effector CTL are not maintained because they require antigenic stimulation to be created, and they are short-lived. Since effector cells are required to combat the infection, the virus will be able to grow without inhibition at the beginning of the secondary challenge. Thus, protection against secondary challenge depends mainly on the amount of time required for the CTL to migrate to the focus of infection and to differentiate into effector cells. In our model, this is captured in the parameter cq, the rate of differentiation into effector cells. Fig 3.5 shows the effect of increased memory CTL abundance on the size of the peak virus load on secondary challenge, assuming different rates of effector cell production (cq). Increased memory CTL levels are only protective if effector function is produced sufficiently fast (large cq) once the pathogen has entered the host. Strikingly, if there is a longer time delay in the production of effector function (small cq), increasing the abundance of memory CTL even by four orders of magnitude does not lead to a significant reduction of the peak virus load and thus of clinical symptoms (Fig 3.5).

3.5.2 Protection During Viral Persistence

If the virus persists after primary infection (above or below detection limit), memory CTL will continue to become activated and to differentiate into effector cells. Therefore, effector responses are maintained in the long-term. The size of the memory population upon rechallenge is given by (2.16), and the size of the effector population by (2.17). These are the equilibrium expressions. At equilibrium, the birth rate equals the death rate, and the overall growth rate of the virus is zero. Upon secondary challenge, a relatively small amount of the same virus is added to the system that is at equilibrium. Therefore, the virus from the secondary challenge cannot grow according to the model. Consequently, on secondary challenge, the host is always protected against clinical symptoms in this scenario. However, this protection may be compromised if reinfection occurs with a virus strain that has mutated to become competitively superior to the immunizing strain, e.g., due to a faster rate of replication (Fig 3.6). In this case, the invading virus variant will have a positive initial growth rate and can therefore replicate up to a peak. The level of the peak virus load increases with increasing competitive superiority of the mutant compared to the wild-type (Fig 3.6).

3.6 Protection Against Rechallenge: Experimental Studies

The mathematical analysis described above points toward the importance of the initial CTL kinetics in protection against rechallenge. It is key that effectors are present at the site of infection as soon as possible upon challenge. If antigen persists this is easily achieved. If antigen does not persist, it may depend on the exact circumstances of infection whether the memory CTL can prevent symptoms. In the experimental literature, there has been considerable controversy about the exact nature and protectiveness of CTL memory. In particular, the role of persisting antigen for maintaining CTL memory and ensuring an efficient secondary response is under intense debate [Beverley (1990); Bruno et al. (1995); Gray and Matzinger (1991); Hou et al. (1994); Jamieson and Ahmed (1989); Kundig et al. (1996a); Kundig et al. (1996b); Lau et al. (1994); Mullbacher (1994); Oehen et al. (1992)]. Experimental results indicate that maintenance of memory CTL is antigen-independent, and

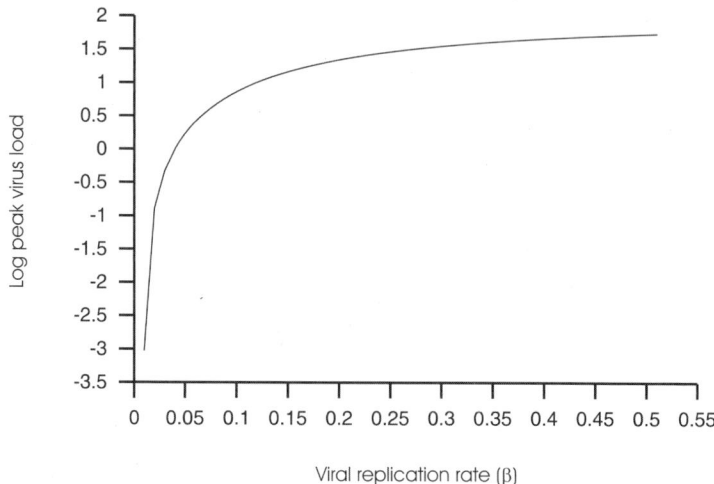

Fig. 3.6. Protection against secondary challenge when the first virus persists in the host. The graph is derived from model (2.12–2.13). If antigen persists the host is protected if reinfection occurs with the same virus strain. However, reinfection with a competitively superior strain results in an increase in peak virus load and thus in reduced protection against secondary challenge. The higher the relative fitness (replication rate) of the second strain, the higher the peak virus load upon infection. Baseline parameters were chosen as follows: $\lambda = 10$, $d = 0.1$, $\beta = 0.001$, $a = 0.5$, $p = 1$, $c = 0.1$, $b = 0.001$, $h = 0.1$.

that the efficacy of the secondary response may or may not require persistent antigen, depending on the kinetics of effector cell production [Ehl et al. (1997); Kundig et al. (1996a); Kundig et al. (1996b)]. This has been shown with LCMV and VSV infection in mice. If the secondary infection is intravenous, then protection seems to be independent of the persistence of antigen [Kundig et al. (1996a); Kundig et al. (1996b)]. In this case, the virus population directly encounters the memory CTL, which leads to instant CTL activation and thus termination of the infection before the appearance of clinical symptoms. On the other hand, protection against peripheral infection appears to be dependent on the persistence of antigen [Kundig et al. (1996a); Kundig et al. (1996b)]. This is because antigen persistence induces the expression of relevant markers on the surface of CTL, such as LFA-1 and VLA-4 [Andersson et al. (1994); Zimmerman et al. (1996)]. This ensures constant recirculation through nonlymphoid tissues, which is required to recognize the invading virus before it has already replicated to high levels. Recently, the effectiveness of protection against secondary challenge was compared directly in an experimental system, immunizing mice both with LCMV and recombinant *Listeria monocytogenes* expressing the nucleoprotein of LCMV [Ochsenbein et al. (1999)]. In contrast to recombinant *Listeria* immunization, antigen was reported to persist at low levels following LCMV immunization. At comparable levels of memory CTL, protection against secondary LCMV challenge was significantly more efficient for LCMV-induced memory CTL compared to recombinant *Listeria*-induced memory CTL. Thus, in agreement with our theoretical results, experiments show that CTL memory tends to be more protective if the time window between invasion of the pathogen and induction of antiviral CTL-effector activity is short, as a result of persistent antigen that keeps the memory CTL activated and circulating through the body.

3.7 Summary

This chapter has described how a relatively simple mathematical model can be used to gain important insights into the importance of CTL memory in viral infections. The model gave rise to the new finding that in addition to protection against secondary challenges, memory may be required for the resolution of primary infection. Impairment of memory could be a decisive defect in CTL responses that could account for persistent and productive infections. This theory will be applied to HIV and HCV infection in Chapters 11 and 12. In addition to these new insights, we have demonstrated how the model has been used to gain some basic quantitative insights into the factors that determine whether memory CTL are protective against secondary challenges. Finally, we have shown how the models have been applied to experimental data in order to address the hypotheses that have been generated.

4
CD4 T Cell Help

The two main branches of the adaptive immune system that directly fight infections are the CTL and the antibodies. The adaptive immune system, however, contains another branch: the CD4 T helper cells. Just as the CTL, they bear a T cell receptor that can specifically recognize antigen. Instead of the CD8 surface receptor that is found on CTL, the helper cells bear the CD4 receptor. Upon antigenic stimulation, the helper cells become activated, undergo clonal expansion, and differentiate into memory cells. While it has been documented that CD4 T cells can have direct antiviral activity [Doherty et al. (1997)], this is thought to be not their main function. As their name implies, their prevalent function is to help the CTL and antibody responses to develop and to fight the infection. This chapter investigates the dynamics of CD4 T cell help in the context of CTL responses.

When does CD4 T cell help influence the CTL (Fig 4.1)? The response to an infection can roughly be divided into two phases. (i) During the acute phase, the naive CTL expand, differentiate into effector cells, and fight the virus. This is called the primary response. It can result in the resolution of the infection. (ii) At this stage, we enter the memory phase. That is, the effector cells differentiate into memory CTL that can persist at elevated levels in the long-term. Upon further antigenic stimulation, the memory cells can become reactivated, proliferate, and differentiate into effector cells again that fight the virus. This is called the secondary response. These effectors can in turn differentiate into more memory cells. If an infection is not cleared but persists in the host, this cycle of memory cell stimulation, proliferation, and differentiation is repeated continuously. Experimental data have shown that the primary response develops normally in the absence of help [Borrow et al. (1998); Janssen et al. (2003); Shedlock and Shen (2003); Sun and Bevan (2003); Thomsen et al. (1996); Thomsen et al. (1998)]. It is therefore called helper-independent. On the other hand, efficient restimulation of memory CTL appears to require CD4 cell help [Borrow et al. (1998); Janssen et al. (2003); Shedlock and Shen (2003); Sun and Bevan (2003); Thomsen et al. (1996); Thomsen et al. (1998)]. The mechanism underlying this observation is

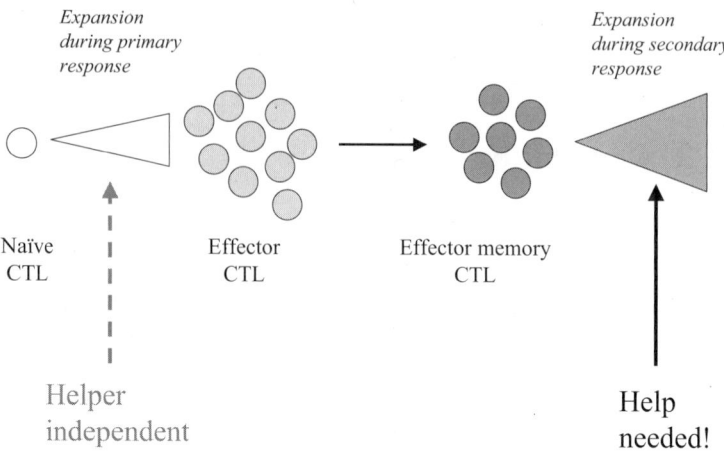

Fig. 4.1. Help is required for the expansion of memory CTL upon restimulation, but help is not required for CTL expansion during the primary response.

unclear [Shedlock and Shen (2003); Sun and Bevan (2003); Sun et al. (2004); Thomsen et al. (1996)]. Some data suggest that the interactions between CD4 cells and CTL is a one time event that occurs early during primary infection; this results in the programming of the CTL, such that the memory cells can subsequently react to further antigenic stimulation. Other data indicate that the interactions between CD4 helper cells and CTL is required on a continuous basis for memory CTL to divide and to expand. Irrespective of the details, however, it appears that after a helper-independent primary CTL response, further expansion of the CTL needs CD4 cell help.

The exact mechanisms by which CD4 cells deliver help to the CTL are still unclear. Two main hypotheses can be found in the literature (Fig 4.2) [Ridge et al. (1998); Schoenberger et al. (1998)]. Traditionally, the main role of CD4 T cells was thought to be the secretion of cytokines upon contact with antigen. These cytokines, especially IL-2, are thought to stimulate CTL, promoting their expansion and maintenance. This mechanism will be referred to as the *classical pathway* (Fig 4.2a). However, recently, it has become clear that a more complicated pathway might be involved in the generation of help for CTL responses [Ridge et al. (1998); Schoenberger et al. (1998)]. Experiments suggest that CD4 T cells can specifically interact with antigen presenting cells (APC), thereby activating them. Activated APCs in turn can specifically interact with CTL, delivering help in the form of costimulatory signals. We will refer to this mechanism as the CD4-APC-CTL pathway (Fig 4.2b).

In this chapter, we derive mathematical models that describe both pathways of help. This will form the basis for the chapters that examine the dynamics of virus-induced helper cell impairment (Chapter 11) and the treatment of immunosuppressive infections (Chapter 12). We will use these models to com-

(i) Classical pathway

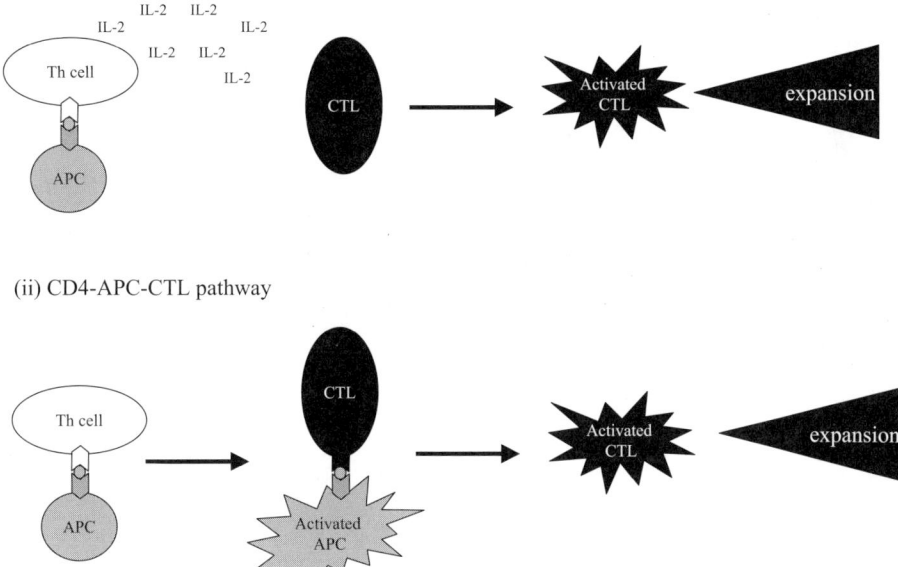

(ii) CD4-APC-CTL pathway

Fig. 4.2. Two mechanisms by which CD4 T helper cells stimulate CTL. (i) According to the classical pathway, the interaction between helper cells and APC induces the helper cells to secrete cytokines, such as IL-2. These cytokines induce CTL activation and expansion. (ii) According to the CD4-APC-CTL pathway, the activation process involves more steps. First the CD4 helper cells specifically interact with and activate the APCs. The activated APCs in turn specifically interact with the CTL. This results in CTL activation and expansion.

pare the two pathways of help, and to determine their ability to contribute to virus control. These consideration will only concern chronic infections because help is only required for the stimulation of memory CTL. We will then include a helper independent primary CTL response and compare the infection dynamics in wild-type and helper deficient hosts.

The CD4-APC-CTL Pathway

According to the CD4-APC-CTL pathway, CD4 T cells specifically activate APCs, and activated APCs then specifically interact with CTL, promoting their expansion. This is mediated by receptor-ligand interactions, such as CD40-CD40L or B7-CD28 [Andreasen et al. (2000); Borrow et al. (1998); Thomsen et al. (1998)]. Thus, the CD4-APC-CTL pathway can be represented by the following reaction scheme:

$$Th + A \underset{k_2}{\overset{k_1}{\rightleftharpoons}} ThA$$
$$ThA \overset{k_3}{\rightarrow} Th + A^*$$
$$CD8 + A^* \underset{k_5}{\overset{k_4}{\rightleftharpoons}} CD8A^* \qquad (4.1)$$
$$CD8A^* \overset{k_6}{\rightarrow} (n+1)\,CD8 + A^*$$
$$A^* \overset{k_7}{\rightarrow} A$$

where $CD8$ stands for CTL, Th stands for helper cells, A stands for APC, and a star represents an activated state. Every CTL that binds to an activated APC and that itself will get activated will through proliferation give rise to n new CTL. Note that in assuming this we have aggregated CTL activation and proliferation in one process. The k_i are reaction constants. It is assumed that the number of APCs is small and that this limits the reactions. The kinetics of these reactions are given by:

$$\frac{d[A]}{dt} = -k_1[A][Th] + k_2[ThA] + k_7[A^*]$$

$$\frac{d[A^*]}{dt} = k_3[ThA] - k_4[A^*][CD8] + (k_5 + k_6)[CD8A^*] - k_7[A^*]$$

$$\frac{d[Th]}{dt} = (k_2 + k_3)[ThA] - k_1[A][Th]$$

$$\frac{d[CD8]}{dt} = -k_4[A^*][CD8] + (k_5 + (n+1)k_6)[CD8A^*]$$

$$\frac{d[ThA]}{dt} = k_1[A][Th] - (k_2 + k_3)[ThA]$$

$$\frac{d[CD8A^*]}{dt} = k_4[A^*][CD8] - (k_5 + k_6)[CD8A^*]$$

where the square brackets denote concentrations of cell types or complexes of cells. We next assume that the kinetics of complexes are fast compared to the other reactions. This amounts to the assumptions that $k_2 + k_3 \gg k_1$ and $k_5 + k_6 \gg k_4$. Under these assumptions both complexes go to their quasi steady state and the concentration will approximately be $[ThA] = \frac{k_1}{k_2 + k_3}[Th][A]$ and $[CD8A^*] = \frac{k_4}{k_5 + k_6}[A^*][CD8]$. This reduces the kinetics to:

$$\frac{d[A]}{dt} = -\frac{k_1 k_3 x}{k_2 + k_3}[A] + k_7[A^*]$$

$$\frac{d[A^*]}{dt} = \frac{k_1 k_3 x}{k_2 + k_3}[A] - k_7[A^*]$$

$$\frac{d[CD8]}{dt} = n\frac{k_4 k_6}{k_5 + k_6}[A^*][CD8]$$

Note that the concentration of free helper cells is constant under these assumptions. Let the total number of helper cells be given by $x = [Th] + [ThA]$.

Under the assumptions made the number of helper cells in complexes is small compared to the number of free helper cells. We can therefore assume that $[Th] = x$. The total number of APCs is given by $y = [A] + [A^*] + \frac{k_1 x}{k_2 + k_3}[A] + \frac{k_4}{k_5 + k_6}[A^*][CD8]$. Note that the number of free APC's $[A] + [A^*]$ is constant and the ratio of free activated to free non-activated APC's will equilibrate. At equilibrium we find:

$$[A^*] = \frac{k_1 k_3 xy}{k_7(k_2 + k_3) + k_1 k_3 x(1 + \frac{k_4}{k_5 + k_6}[CD8]) + k_1 k_7 x}$$

$$= \frac{\varepsilon xy}{1 + \varepsilon x(1 + f_c z) + f_h x}$$

where $\varepsilon = \frac{k_1 k_3}{k_7(k_2 + k_3)}$ is the net reaction constant of APC activation $c = n\frac{k_4 k_6}{k_5 + k_6}$ is the net reaction constant of CTL activation $f_h = \frac{k_1}{k_2 + k_3}$ is a proportionality constant for the Th-APC complex, and $f_c = \frac{k_4}{k_5 + k_6}$ is a proportionality constant for the CTL-APC complex.

We denote the number of CTL by $z = [CD8] + [CD8A^*]$. Under the assumption that the number of CTL in complexes is negligible compared to the total number, we can assume $[CD8] = z$. The proliferation of these cells is described by:

$$\frac{c\,\varepsilon\,xyz}{1 + \varepsilon x(1 + f_c z) + f_h x} \tag{4.2}$$

This is a form of the proliferation function that generalises several functions that have been described previously. For instance, by assuming that the amount of help x is constant we find:

$$\frac{\alpha y}{\beta + z} z \tag{4.3}$$

where $\alpha = c/f_c$, $\beta = 1/(\varepsilon x f_c) + (\varepsilon + f_h)/(\varepsilon f_c)$. This proliferation function has been derived by [Borghans et al. (1999); De Boer and Perelson (1994); Fishman and Perelson (1993)] (see also Chapter 2). Another possible assumption is that the help is vanishingly small, which amounts to linearising the proliferation function in x in the neighbourhood of $x = 0$, which results in:

$$c\varepsilon xyz \tag{4.4}$$

which is arguably the simplest possible functional form (Nowak and Bangham 1996, Wodarz and Nowak 1999). In the derivation of the general proliferation function it was assumed that $k_2 + k_3 \gg k_1$ and $k_5 + k_6 \gg k_4$. Hence it is natural to assume that f_h and f_c can be ignored, which results in:

$$\frac{c\,\varepsilon\,xyz}{1 + \varepsilon x}. \tag{4.5}$$

This functional form has also been used to describe CTL dynamics [Nowak and Bangham (1996); Wodarz and Nowak (1999)] (see also Chapter 2). In the following, we will use this particular functional form. The rate of CTL activation and proliferation is thus a saturating function of the number of helper cells with $1/\varepsilon$ as the half saturation constant. In the presence of large amounts of help, the rate of CTL proliferation approximates cyz, where the parameter c denotes the CTL responsiveness. This parameter includes a variety of factors that influence the ability of the CTL to respond to antigen, determined by the T cell receptor and the MHC type of the host. The level of help is given by the number of helper cells x, as well as the efficacy of those helper cells ε. The higher the efficacy of help ε the lower the number of helper cells x required to induce maximal stimulation of the CTL response.

If we next assume that CD8 cells die with a rate bz we get the kinetics

$$\frac{dz}{dt} = \frac{c\varepsilon xyz}{1+\varepsilon x} - bz \qquad (4.6)$$

In this model, the CTL response becomes established if $\varepsilon x > \frac{b}{(c\tilde{y}-b)}$, where $\tilde{y}$ denotes virus load in the absence of a CTL response. In other words, the CTL response only becomes established if the amount of CD4 cell help lies above a critical threshold. This result makes sense in that an immune response is wasted when triggered by too low concentrations of antigen. Equilibrium virus load in the presence of CTL is given by $\hat{y} = \frac{b(1+\varepsilon \hat{x})}{c\varepsilon \hat{x}}$. As shown in Fig 4.3, an increase in the amount of help reduces virus load down to an asymptote, at which the helper-induced stimulation of the CTL response has reached its maximum. At the asymptote $\hat{y} = b/c$.

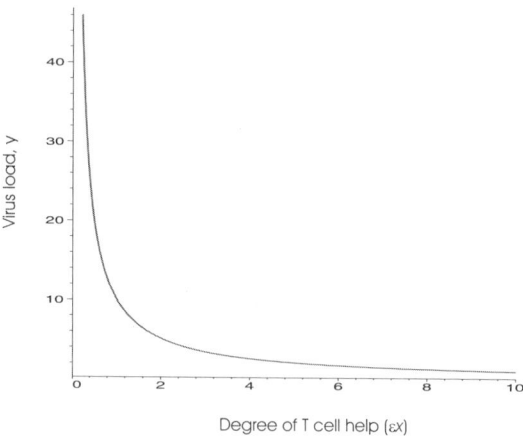

Fig. 4.3. Effect of CD4 cell help on virus load, according to equation (4.6). Virus load decreases asymptotically with increasing degrees of help. Virus load approaches a value of b/c for high degrees of help. Parameters were chosen as follows: $c = 1$, $b = 0.1$, $\epsilon = 0.01$.

4 CD4 T Cell Help

The equilibrium viral load is determined by other immunological processes. Within the framework presented here we can model the proliferation of CD4 T helper cells in response to activation by APCs. If we assume that helper T cell proliferation is proportional to the amount of helper T cells in complexes we find that helper T cell proliferation is given by: ϕxy, where ϕ is the proliferation rate. If the death rate of helper T cells is given by ψ we can write for the helper T cell dynamics:

$$\frac{dx}{dt} = \phi xy - \psi x \qquad (4.7)$$

The Classical Pathway

For the classical pathway of T cell proliferation it is assumed that CTL require contact with the antigen and exposure to cytokines, denoted I, in order to proliferate. The cytokines are released by CD4 cells upon contact with antigen. This amounts to the following reaction scheme:

$$\begin{aligned} Th + A &\underset{k_1}{\overset{k_2}{\rightleftharpoons}} ThA \\ ThA &\overset{k_3}{\to} Th + A + I \\ CD8 + A &\underset{k_4}{\overset{k_5}{\rightleftharpoons}} CD8A \\ CD8A + I &\overset{k_6}{\to} (n+1)\,CD8 + A + I \\ I &\overset{k_9}{\to} . \end{aligned} \qquad (4.8)$$

with kinetics:

$$\frac{d[A]}{dt} = -k_1[Th][A] + (k_2 + k_3)[ThA] - k_4[CD8][A] + (k_5 + k_6[I])[CD8A]$$

$$\frac{d[Th]}{dt} = -k_1[Th][A] + (k_2 + k_3)[ThA]$$

$$\frac{d[CD8]}{dt} = -k_4[CD8][A] + (k_5 + (n+1)k_6[I])[CD8A]$$

$$\frac{d[I]}{dt} = k_3[Th][A] - k_9[I]$$

$$\frac{d[CD8A]}{dt} = k_4[CD8][A] - (k_5 + k_6[I])[CD8A]$$

$$\frac{d[ThA]}{dt} = k_1[Th][A] - (k_2 + k_3)[ThA]$$

Under assumption that $k_2 + k_3 \gg k_1$ and $k_5 + k_6[I] \gg k_4$ we can assume that the complexes CD8A and ThA go to their quasi steady states: $[CD8A] = \frac{k_4}{k_5 + k_6[I]}[CD8][A]$, and $[ThA] = \frac{k_1}{k_2 + k_3}[Th][A]$. Under these assumptions the free Th, denoted x, and APC, denoted y, populations are constant (note that this tacitly assumes the number of Ths and APCs in complexes is negligible compared to the number of free cells). The kinetics simplify to:

$$\frac{d[CD8]}{dt} = \frac{nk_4 k_6 [I]}{k_5 + k_6 [I]} [CD8] y$$

$$\frac{d[I]}{dt} = k_3 xy - k_9 [I]$$

If, in addition, we assume that the cytokine concentration goes to equilibrium, and by denoting $[CD8] = z$ we get the proliferation rate:

$$\frac{\gamma x y^2 z}{1 + \eta x y} \tag{4.9}$$

where $\gamma = \frac{nk_4 k_6 k_3}{k_5 k_9}$ and $\eta = \frac{k_6 k_3}{k_5 k_9}$. According to this expression, the efficacy of such a response depends on the amount of antigenic stimulation. If the antigen concentration is large the proliferation function goes to $\frac{\gamma}{\eta} yz$ and the effect of help disappears. In other words the response is likely to be effective at high virus loads. However, for low antigen concentrations we find: $\gamma x y^2 z$. For low help we recover the same expression. Note that this differs from the proliferation function derived before. Under the classical pathway the proliferation is much lower for low amounts of antigen and an immune response will be slow. The T helper cell proliferation will not be affected by this different pathway.

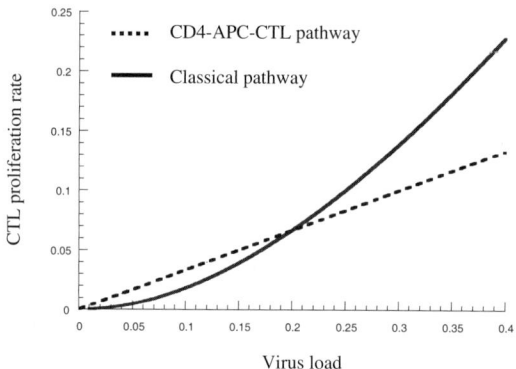

Fig. 4.4. The rate of CTL proliferation as a function of virus load for the two pathways of help, based on equations (4.5) and (4.9). The relationship is linear for the CD4-APC-CTL pathway, while it is not for the classical pathway. The CD4-APC-CTL pathway results in more efficient CTL proliferation at low virus loads. The classical pathway results in more efficient CTL proliferation at high virus loads. Parameters were chosen as follows: $c = 1$, $\epsilon = 0.5$, $\gamma = 2$, $x = 1$, $\eta = 1$.

4.1 Comparison of the Two Helper Pathways

The kinetics of the classical pathway are compared to those of the CD4-APC-CTL pathway in Fig 4.4, plotting the rate of CTL proliferation as a function of antigen concentration. While the rate of CTL proliferation is a linear function of antigen concentration under the CD4-APC-CTL pathway, the relationship is not linear under the classical pathway. The classical pathway induces CTL proliferation more efficiently at high antigen concentrations, while the CD4-APC-CTL pathway induces CTL proliferation more efficiently at low antigen concentrations (Fig 4.4). This observation follows from the derivation of the proliferation responses. The maximum response under the CD4-APC-CTL pathway is cy, while under the classical pathway the maximum response is given by $\gamma/\eta y$. Because $c = \frac{nk_4k_6}{k_5+k_6} < nk_4 = \gamma/\eta$ the maximum response of the classical pathway exceeds that of the CD4-APC-CTL pathway.

4.2 How Does Help Work?

We have considered two possible mechanisms of help in the context of mathematical modeling: the CD4-APC-CTL pathway, and the classical pathway. There is evidence supporting a potential role for both of these mechanisms. Experiments suggest that in the absence of any form of help, CTL become nonresponsive within days following antigenic challenge, not being able to sustain their own proliferation [Deeths et al. (1999)]. However, exogenous administration of IL-2 resulted in stimulation and proliferation of the CTL [Deeths et al. (1999)]. On the other hand, experiments also suggest that help can be administered directly by antigen presenting cells in the absence of CD4 cells. CD4 cells are thought to activate APCs through specific receptor ligand interactions, such as CD40 and CD40L. Crosslinking of CD40 on APCs results in the delivery of help without the need for CD4 cells [Ridge et al. (1998); Schoenberger et al. (1998)]. With these findings in mind, questions arise regarding the exact nature of help and the relative roles of these two pathways.

In this respect, our modeling approach offers the following insights. The models suggest that the two pathways of help might both be required in different situations. At high virus loads, the rate of CTL expansion is faster with the classical pathway compared to the CD4-APC-CTL pathway (Fig 4.4). On the other hand, at low loads, the rate of CTL expansion is slower with the classical pathway compared to the CD4-APC-CTL pathway (Fig 4.4). This could translate to the following situation during the infectious process. The classical pathway might be needed if the CTL response needs to expand quickly in the face of relatively high or increasing virus loads. This can occur, for example, if the host gets infected a second time and the memory CTL have to react to a growing virus population. On the other hand, the CD4-APC-CTL pathway could be needed for the CTL to fight the infection efficiently at low virus loads. This applies to the resolution phase of the infection, when the CTL have already reduced virus load to low levels and need to drive the remaining virus population extinct. If the CD4-APC-CTL pathway was not available, modeling suggests that resolution of the infection is likely to be incomplete: although virus load is initially reduced to low levels, immunological pressure is lost at low loads, resulting in the ability of the virus population to grow back and establish a persistent infection (see discussion in Chapter 3).

These verbal arguments can be supported by a simple model, taking both pathways of T cell help into account. The model is given by the following set of differential equations.

$$\dot{x} = \phi xy - \psi x, \tag{4.10}$$

$$\dot{y} = ry(1 - y/\kappa) - pyz, \tag{4.11}$$

$$\dot{z} = \frac{(c\varepsilon x + k)yz}{(1 + \varepsilon x + k)} + \frac{(\gamma xy^2 z)}{(1 + \eta xy)} - bz. \tag{4.12}$$

The variable x stands for helper cells, y denotes the replicating virus population, and z stands for the CTL response. Virus replication is modeled as a simple density dependent growth process. The helper cell population expands in response to antigenic stimulation, and delivers help to the CTL via both the classical and the CD4-APC-CTL pathway, as derived in this chapter. Fig 4.5 shows simulations of this model for strong and weak helper cell responses. The graphs plot virus growth and the rate of memory CTL expansion in the context of a secondary infection. The figure distinguishes the rate of CTL expansion as a result of the two pathways of help: the classical pathway and the CD4-APC-CTL pathway. If the helper cell response is strong, the outcome of the model is suppression of virus load to low levels (Fig 4.5a). If we assume that extinction occurs below a threshold level of virus load, the CTL response clears the pathogen. At the beginning of the infectious process, the classical pathway induces faster CTL proliferation than the CD4-APC-CTL pathway (Fig 4.5a). However, when virus load starts to decline to lower levels, the CD4-APC-CTL pathway results in faster induction of CTL proliferation that results in the eventual clearance of the virus population (Fig 4.5a). On the other hand, if the helper cell response is weak, the CTL response fails to clear the infection resulting in persistent virus replication (Fig 4.5b). Initially, the classical pathway induces a relatively strong CTL responsiveness (Fig 4.5b). This is because at high loads, CTL proliferation resulting from the classical pathway becomes less dependent on the amount of help. However, immunological pressure is lost at lower virus loads resulting in failure to suppress the infection in the long-term (Fig 4.5b).

In summary, our mathematical models suggest different roles for the two pathways of help: the classical pathway is required to react quickly to high virus loads. Hence, the classical pathway could kickstart the CTL response during the initial phases of the infectious process. On the other hand, the CD4-APC-CTL pathway is required to ensure an efficient and fast response

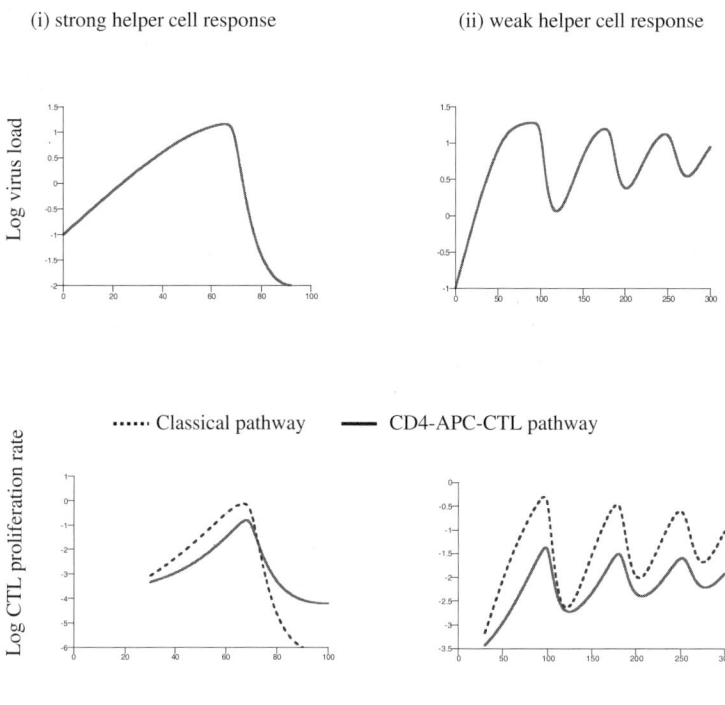

Fig. 4.5. Simulation showing the relevance of the classical pathway and the CD4-APC-CTL pathway in the different stages of the infectious process, according to equations (4.10–4.12). We consider virus growth and the rate of memory CTL expansion in the context of secondary stimulation, because help is only relevant in this context. (i) Simulation assuming a relatively strong helper cell response, resulting in resolution of the infection. Initially, at high virus loads, the classical pathway is more efficient at inducing CTL proliferation. In the clearance phase of the infection, the CD4-APC-CTL pathway is more efficient at maintaining CTL proliferation at low virus load. Note that we assume an extinction threshold for the virus, since the model is deterministic. (ii) Simulation assuming a relatively weak helper cell response. Although initially the classical pathway induces relatively strong degrees of CTL proliferation at high virus loads, immunological pressure is lost at low loads, resulting in failure of the CTL response to resolve the infection. Parameters were chosen as follows: $\psi = 0.05$, $r = 0.1$, $\kappa = 20$, $p = 1$, $c = 0.08$, $\varepsilon = 1$, $\gamma = 0.08$, $\eta = 1$, $b = 0.1$; For (a), $\phi = 0.02$; For (b) $\phi = 0.007$.

at low virus loads. Hence, the CD4-APC-CTL pathway is important to ensure viral clearance following the acute phase of the infection.

4.3 Infection Dynamics in Helper Deficient Hosts

So far, we have investigated the dynamics of CD4 T cell help for CTL responses. Because help is only required for the expansion of memory CTL, and not for the initial expansion of primary CTL, these considerations do not apply to acute or primary infection dynamics. Instead, they apply to the dynamics during the chronic phase of a persistent infection, or to secondary infections. In this section, we investigate infection dynamics in the absence of help. That is, we assume that no CD4 cell help is available in the host. This scenario corresponds to helper deficient mice. In this case, there is an intact primary CTL response, but no long-term memory CTL response. The helper-independent primary CTL response is modeled with the equations that describe programmed CTL proliferation (2.18–2.23). That is, upon antigenic stimulation, the naive CTL become activated and divide a certain number of times. This leads to their differentiation into effector CTL that have antiviral activity. Effector CTL in turn differentiate into memory CTL. Because we assume that no help is available, these memory CTL cannot react and give rise to antiviral activity upon further antigenic stimulation anymore.

In the context of this model, we ask under which conditions such a CTL response can clear the infection, and when persistent infection is established. This will tell us under which conditions help is required for the clearance of infection and when it is not. Experimental data suggest that several infections, such as LCMV and γ-herpes virus, need help for clearance [Borrow et al. (1998); Sarawar et al. (2001); Thomsen et al. (1998)]. Other infections, such as murine infeluenza virus, however, can be cleared in the absence of help [Belz et al. (2002)].

We ask how host and viral parameters influence the ability of an acute (helper-independent) CTL response to clear an infection (Fig 4.6). Regarding CTL parameters, we find an optimal rate of CTL activation (α) and proliferation (r) that maximizes the chance of pathogen clearance. Variation in both parameters produces similar trends (Fig 4.6). As the rate of CTL activation / proliferation is increased from low to high, the degree of CTL-mediated reduction of virus load becomes stronger (i.e. clearance becomes more likely). The reason obviously is that the immune response becomes more effective. As the rate of CTL activation/proliferation is increased further, however, the degree virus load reduction by CTL becomes weaker (i.e. clearance becomes less likely). This is because effector activity is generated fast and the virus can only grow to limited levels. Limited antigenic load reduces the chances to trigger further naive CTL to expand. This results in the generation of overall fewer effector cells and thus in a reduced chance of clearance. Hence, there is a tradeoff between the extent of initial virus growth and the ability of the CTL to clear the infection. If virus growth is stopped too early and virus load does not reach higher levels, acute pathology is reduced, but fewer CTL effectors are generated. This results in a reduced ability to clear. If virus grows to

4.3 Infection Dynamics in Helper Deficient Hosts   67

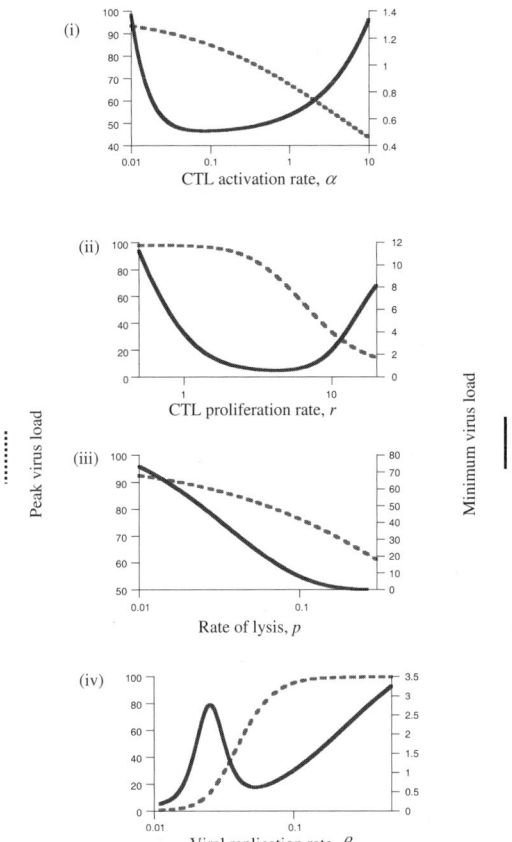

Fig. 4.6. Relationship between host and viral parameters and virus dynamics during acute infection. Among the host parameters, we consider the rate of CTL activation and proliferation (α and r), as well as the rate of CTL-mediated antiviral activity (p). Among the viral parameters, we consider the replication rate (β). Two measures are plotted. The minimum virus load is an indicator of the chance that the infection is cleared. The lower the minimum load during acute infection the higher the chances of clearance. The peak virus load indicates the severity of acute symptoms. Simulations are based on the model of programmed CTL proliferation (2.18–2.23). For explanation, see text. Parameters were chosen as follows: $\lambda = 10$, $d = 0.1$, $\beta = 0.05$, $a = 0.1$, $r = 5$, $\gamma = 1$, $\alpha = 0.01$, $\delta = 0.5$, $p = 0.2$, $\epsilon = 0.001$.

higher levels, stronger pathology is observed, but the infection is more likely to be cleared. This tradeoff between limiting acute infection and ensuring viral clearance is a general property of the mathematical model that describes programmed proliferation (2.18–2.23), and is also discussed in Chapter 2. In contrast to this counterintuitive outcome, the relationship between the rate of antiviral activity p and the chance of clearance is straightforward (Fig 4.6):

the higher the rate of antiviral activity, the higher the chances to achieve virus clearance.

Among viral parameters, the replication kinetics of the virus β significantly determines the chances of virus clearance during the acute phase. We observe an interesting relationship if the viral replication kinetics are changed from low to high (Fig 4.6). (i) At first, an increase in viral replication reduces the chances of viral clearance because the virus spreads faster. This corresponds to the parameter region where the basic reproductive ratio of the virus is close to one and the infection can just about be maintained. (ii) A further increase in the viral replication kinetics, however, changes this relationship. Now, faster virus replication leads to higher chances of clearance. This is because faster spread gives rise quickly to higher levels of antigen. While this increases acute pathology, it also induces more naive CTL to undergo expansion that results in more effectors and more efficient clearance. On the other hand, slower virus spread results in lower levels of antigen. While this reduces acute pathology, it triggers fewer naive CTL into expansion, fewer effectors are generated, and this renders clearance less efficient. The virus can be thought of as sneaking past the CTL response by replicating relatively slowly and thereby providing a weak stimulus. This might be observed with relatively slowly replicating viruses, and has been suggested previously as a reason for the persistence of some hepatitis C virus strains [Bocharov et al. (2004)]. Thus, there is again a similar tradeoff between the efficacy of clearance and the ability of the response to reduce acute symptoms. (iii) Finally, if the viral replication kinetics are increased further, faster virus spread again decreases the chances of virus clearance. This is because in this parameter region, virus spread is sufficiently fast that at the time of CTL triggering, virus load has already attained the highest possible values. Therefore, a further increase in the viral replication kinetics does not change the level of antigenic stimulation, but merely counters the effect of CTL-mediated antiviral activity. This is likely to be observed with relatively fast replicating viruses.

If we would like to determine which viruses can be cleared by CTL in the absence of help, we have to concentrate on differences in viral parameters. According to the above discussion, the viral replication rate is the most important viral variable that can determine whether a helper-independent acute CTL response can clear the infection. As discussed above, the relationship between the viral replication kinetics and the ability of a single round of proliferation to clear the infection is complex. For relatively slowly replicating viruses it is possible that an increase in the rate of viral spread increases the chances of viral clearance because CTL receive a higher antigenic stimulus. On the other hand, in the parameter region where viral replication is generally faster, an increase in the rate of viral spread always increases the chance of viral persistence. Therefore, when comparing viruses that are characterized by different replication rates in general, it is difficult to predict how replication kinetics correlate with the need for help to resolve the infection.

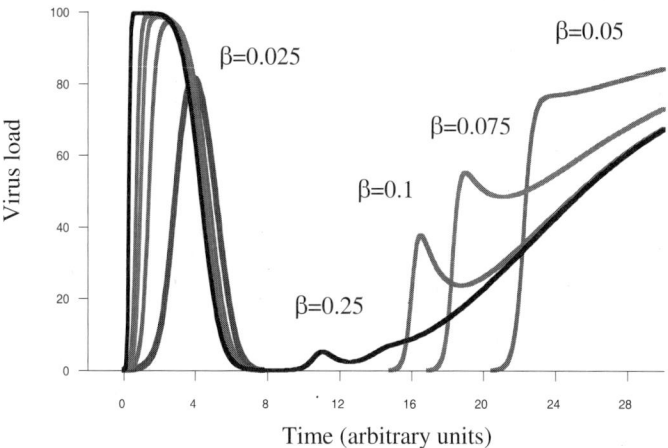

Fig. 4.7. Simulation of virus dynamics in helper deficient hosts. The simulation is based on model (2.18–2.23). We assume that one round of programmed CTL proliferation is induced in helper deficient hosts, which results in the generation of memory cells. The memory cells cannot, however, be reactivated anymore. The dynamics depend on the rates of viral replication. We compare different viruses that are generally fast replicating, but that differ in their exact replication kinetics β. This could correspond to LCMV infection. Virus load grows to a peak and is then downregulated by the CTL response. Because memory cells cannot become reactivated, however, the virus population can grow back from low levels. The lower the rate of viral replication, the longer it takes for the virus to grow back, and the longer the duration of virus control. Fast virus spread correlates with a fast resurgence of virus load. Parameters were chosen as follows: $\lambda = 10$, $d = 0.1$, $a = 0.1$, $r = 1$, $\gamma = 0.1$, $\alpha = 1$, $\delta = 0.5$, $p = 0.4$, $\epsilon = 0.001$.

The easiest scenario is given by the comparison of viruses that differ in their rate of spread but that all replicate at a relatively fast rate. In this parameter region the model suggests that increased viral replication kinetics correlate with an inability of CTL to clear the infection in the absence of help. Moreover, the model suggests the following dynamics in helper deficient hosts (Fig 4.7). The CTL response initially reduces virus load to low or undetectable levels. This is because the acute response does not depend on help. If the viral replication rate is relatively slow, this results in clearance or long-term control. Otherwise, this is followed by a resurge of the virus population after a given period of time, which correlates with the lack of memory CTL responses. The time it takes for the virus to resurge depends on the rate of viral replication. The faster the replication rate, the quicker the virus population grows back (virus load is reduced to a lesser degree and can subsequently grow faster in the absence of memory responses).

Model predictions regarding the dynamics in the absence of help fit well with experimental data from virus infections in helper deficient mice. LCMV is a good case study [Lehmann-Grube (1971)]. It is a relatively fast replicating virus, but comes in different strains characterized by different replication kinetics [Thomsen et al. (2000)]. As suggested by the model, if the replication rate of the virus is slower, such as with LCMV Armstrong, absence of help results in the clearance of the infection [Ahmed et al. (1988)]. Faster replicating strains of LCMV (such as Traub), however, result in persistent infection [Christensen et al. (2001); Planz et al. (1997); Thomsen et al. (1996); Thomsen et al. (2000)]. In accord with the model, the experimental data show that the acute CTL response initially reduces virus load to low levels, and that the virus population subsequently grows back [Thomsen et al. (1996)]. In addition, virus resurgence is observed earlier with faster replicating viruses [Thomsen et al. (1996)].

4.4 Summary

In this chapter, we derived mathematical models that describe the delivery of help by CD4 T cells for CTL responses. We considered two models of help: the classical pathway that assumes that CD4 T cells secrete cytokines that stimulate the CTL; and the so called CD4-APC-CTL pathway that assumes that CD4 T cells need to activate APC, and the activated APCs in turn stimulate the CTL. Comparison of the two pathways showed that the classical pathway is more efficient at driving CTL proliferation at high virus loads, while the CD4-APC-CTL pathway is more efficient at driving CTL proliferation at low virus loads. Therefore, these two pathways of help might be important in different phases of an infection. When the virus population grows to high levels, the classical pathway can make sure that the CTL can react efficiently. When the virus load is already on its way down, the CD4-APC-CTL pathway makes sure that immunological pressure is maintained at low virus load, and that the infection is cleared. We further investigated infection dynamics in helper deficient hosts and determined the conditions when a helper-independent acute CTL response can result in clearance of infection, and when it fails. If help is required for clearance, then the virus population is temporarily controlled in helper deficient hosts, followed by a resurgence of virus load to high levels. The rate of virus resurgence depends on the viral replication rate.

5
Immunodominance

The chapters so far have investigated the dynamics between CTL and viral infections, assuming that a single population of CTL fights a population of viruses. The situation is, however, more complex than this (Fig 5.1). The immune system can generate a huge diversity of T cells that are specific to different antigens. A given virus consists of many parts that can all be potentially recognized by CTL. These are called viral epitopes. Therefore, several different CTL can exist that can recognize different epitopes of the same virus. Each of these CTL can become activated and undergo clonal expansion upon infection. Consequently there is not a single CTL response against an infection. Instead, there are multiple CTL responses or clones that fight the same virus population.

When CTL responses against an infection are quantified *in vivo* over time, several CTL clones directed against different epitopes are measured. A hierarchy exists among the CTL responses, and the relative abundance of the different CTL is referred to as *immunodominance* [Yewdell and Bennink (1999)]. In some cases, it is possible that there is a single immunodominant response, and that all other responses are either absent or present at very low levels. This is referred to as a *narrow response*. In other scenarios, many CTL clones coexist at relatively high levels, and this is referred to as a *broad response*.

This gives rise to the question of how immunodominance comes about, and what determines the pattern of immunodominance observed *in vivo* [Adorini et al. (1988); Bergmann et al. (1999); Brehm et al. (2002); Chen et al. (2000); Goulder et al. (1997b); Wherry et al. (2003a); Yewdell and Bennink (1999)]. Mathematical models have played a useful role in this respect. They suggest that a key concept that can explain immunodominance is the competition between the different CTL clones for antigenic stimulation by the virus population [Nowak (1996); Nowak et al. (1995a)]. Each CTL clone requires antigenic stimulation in order to expand. The CTL clones are likely to differ in the efficiency with which they respond to antigen (i.e. in their responsiveness). A more responsive CTL clone can become stimulated by lower levels of antigen and can suppress virus load to lower levels than a less responsive

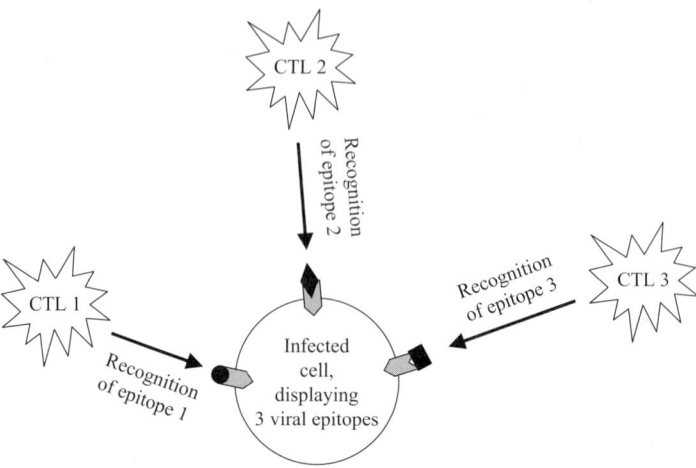

Fig. 5.1. Schematic diagram that illustrates the concept of multiple CTL clones directed against different epitopes of the same virus. An infected cell is depicted. It displays parts of the virus on its cell surface. Different parts of the virus, called epitopes, are displayed. Each of these epitopes can elicit a different CTL clone that is specific for the particular epitope. Therefore, a single virus can induce multiple CTL clones to respond, directed against different parts/epitopes of the same virus.

CTL clone. Imagine the existence of two CTL clones, CTL1 and CTL2. CTL1 can suppress virus load to levels that are too low to stimulate CTL2. In this case, CTL1 will expand and will be maintained, while CTL2 will go extinct, or persist at very low levels. In other words, CTL1 outcompetes CTL2, and CTL1 is said to be immunodominant. This chapter will review mathematical models that analyze these competition dynamics between different CTL that are directed against different epitopes of the same virus population, and discuss how this influences the pattern of immunodominance.

5.1 A Mathematical Model for Multiple CTL Clones

We will consider an extension of the basic mathematical model of CTL responses that distinguishes between memory CTL and CTL effectors (2.12–2.13). The properties of this model are reviewed in Chapter 3. Instead of considering a single CTL response, we now assume the existence of multiple CTL clones ($i = 1..n$) that are directed against different epitopes of the same virus population [Wodarz and Nowak (2000a)]. Therefore, the population of memory CTL is denoted by w_i, and the population of CTL effectors is denoted by z_i. The model is given by the following set of ordinary differential equations.

5.1 A Mathematical Model for Multiple CTL Clones

$$\dot{x} = \lambda - dx - \beta xy, \quad (5.1)$$

$$\dot{y} = \beta xy - ay - y\sum_{i=1}^{n} p_i z_i, \quad (5.2)$$

$$\dot{w}_i = c_i y w_i (1-q) - b w_i, \quad (5.3)$$

$$\dot{z}_i = c_i q y w_i - h z_i. \quad (5.4)$$

We assume that the CTL clones are characterized by different responsiveness to the antigen c_i and by different rates of target cell killing p_i. For simplicity we assume that the CTL responsiveness is positively correlated with the rate of target cell killing, since both parameters are determined by recognition of antigen in conjunction with MHC. However, the results do not depend on this assumption. In such a scenario, the CTL clones are essentially in competition with each other. Competitive ability correlates with the CTL responsiveness of the epitope c_i. The CTL clone directed at the epitope with the largest c_i is the most superior competitor. The outcome of these competitive interactions depends the life span of the CTL response in the absence of antigen, i.e. on the parameter $1/b$ (Fig 5.2).

According to equilibrium analysis of the model, only the clone with the highest CTL responsiveness c_i can survive at significant levels. Mathematically speaking, all other CTL clones go extinct, although in practical terms, the spatial environment of the immune system could result in persistence of these clones at low levels. The reason is that the competitively superior CTL clone reduces virus load to levels too low to stimulate the weaker CTL clones. This argument can be expressed mathematically as follows. Let us rank the CTL clones according to their competitive ability, expressed in the value of c_i: $c_1 > c_2 > c_3 > ... > c_n$. If virus load y has equilibrated, w_i declines with an exponential rate that is described by

$$w_i(t) = e^{-b(1-\frac{c_i}{c_1})t}.$$

Only the most superior competitor survives. In numerical simulations, this result will be obtained if the life span of the CTLp response in the absence of antigen $(1/b)$ is short (Fig 5.2a). A short life span of the CTL in turn correlates with weak virus control (see Chapters 2 and 3). Thus, lack of efficient CTL-mediated control of the infection correlates with the presence of an immunodominant CTL clone (Fig 5.2a).

On the other hand, if the life span of the CTL memory response in the absence of antigen is long (high $1/b$), the dynamics between virus replication and the CTL response converge to a quasiequilibrium. Similar to the basic CTL memory model (2.12–2.13), virus load at the quasi equilibrium $(y^\wedge)$ is higher than virus load at the true equilibrium (y^*), i.e. $y^\wedge = \alpha y^*$, where $\alpha > 1$. Now, coexistence of multiple CTL clones directed against different epitopes is possible (Fig 5.2b). This is because the competitively superior CTL clone does not reduce virus load down to the true equilibrium, but only

(i) CTLp short lived in the absence of antigen

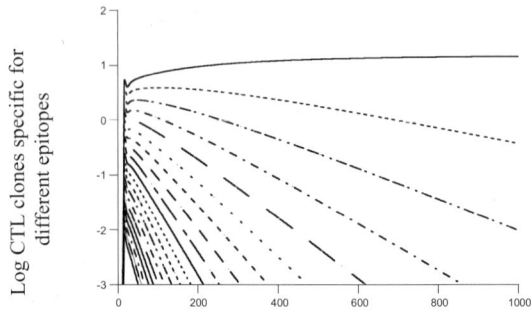

(ii) CTLp long lived in the absence of antigen

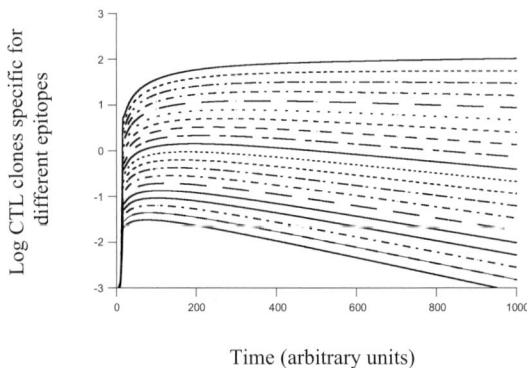

Time (arbitrary units)

Fig. 5.2. Relationship between the longevity of the CTL response in the absence of antigen and the clonal composition of the CTL response. (i) If memory CTL decay at a fast rate in the absence of antigen (large b), most CTL clones are outcompeted by the most immunogenic clone, resulting in immunodominance. This outcome correlates with poor virus control. (ii) If memory CTL decay at a slow rate in the absence of antigen (small b), we observe coexistence of multiple CTL clones directed against different epitopes. This outcome correlates with efficient virus control. Simulations are based on equations (5.1–5.4) Baseline parameters were chosen as follows: $n = 20$, $\lambda = 10$, $d = 0.1$, $\beta = 0.01$, $a = 0.5$, $q = 0.1$, $h = 1$, $c_i = 1 + 0.1i$, $p_i = c_i$. For (i), $b = 0.1$; for (ii) $b = 0.007$.

to the quasiequilibrum, where virus load may still be sufficient to stimulate the weaker CTL clones. We can define this more precisely by ranking the CTL clones against different epitopes according to their competitive ability, i.e. $c_1 > c_2 > c_3 > > c_n$. If the life span of the memory CTL is long, then the dynamics of the CTL clones directed against the different epitopes can be described by

$$w_i(t) = e^{-bt(1-\alpha \frac{c_i}{c_1})}.$$

From this it follows that the CTL clone directed against epitope i w_i persists during the quasiequilibrium if $c_i \alpha > c_1$. In summary, in the presence of a long life span of memory CTL and thus efficient long-term virus control, the dynamics do not lead to one immunodominant CTL clone, but to a broad CTL response directed against multiple epitopes.

5.2 Theory and Data

The theory discussed here presents a simple explanation for different patterns of immunodominance observed in data. The model suggests that weak virus control, caused by a short life span of memory CTL, correlates with the dominance of a single CTL clone. That is, we observe a narrow response. On the other hand, efficient virus control, due to a long life span of memory CTL, is associated with broad CTL response where multiple CTL clones, directed against different epitopes, coexist with each other.

These predictions are supported by data from HIV infected individuals. Long-term nonprogressors had strong CTL responses, low viral load, and a stable CD4 T cell count of > 500 cells per μl blood fifteen years after infection [Harrer et al. (1996a); Harrer et al. (1996b)]. There was a broad CTL response directed against multiple epitopes and maintained at high levels despite the low level of viraemia [Harrer et al. (1996a); Harrer et al. (1996b)]. This supports the prediction that broad CTL responses can be associated with efficient virus control. On the other hand, faster progressing patients, characterized by weaker immunity and less efficient virus control, are characterized by narrow CTL responses [Borrow et al. (1997)], again consistent with model predictions.

Apart from these basic patterns, both theory and experiments have identified other factors that can influence the pattern of immunodominance in viral infections, and that have not been included in the simple model discussed above (5.1–5.4). For example, virus evolution towards antigenic escape in HIV infection can broaden the CTL response [Gupta and Anderson (1999); Nowak (1996); Nowak et al. (1995a)]. Several variants of an epitope exist in the virus population, and the CTL responses against these mutants can coexist. Antigenic escape and heterogeneity usually correlate with disease progression and loss of immunological control in HIV infection [Goulder et al. (1997a); Klenerman et al. (1995); McMichael et al. (1995); McMichael and Phillips (1997); Phillips et al. (1991); Price et al. (1997b)]. Thus, a broad CTL response can be observed in the absence of good virus control, if loss of virus control is caused by the evolution of antigenic heterogeneity, and not by a weak CTL response itself. On the other hand, a narrow CTL response can also arise in specialized circumstances that have not been taken into account in the model.

In Human T lymphocyte virus type 1 (HTLV-1), a single dominant CTL response directed against the TAX protein is observed [Bangham (1999); Goon et al. (2004)]. While this could result from an inefficient CTL response that fails to control the infection at low levels, another explanation is that TAX is the only antigen exposed to CTL, and that infected cells are killed before any other viral proteins can be presented on the cell surface. In addition, the presence of specific viral antigens that are highly immunogenic compared to alternative viral antigens could result in a narrow CTL response, even if it is efficient and can control viral replication in the long-term.

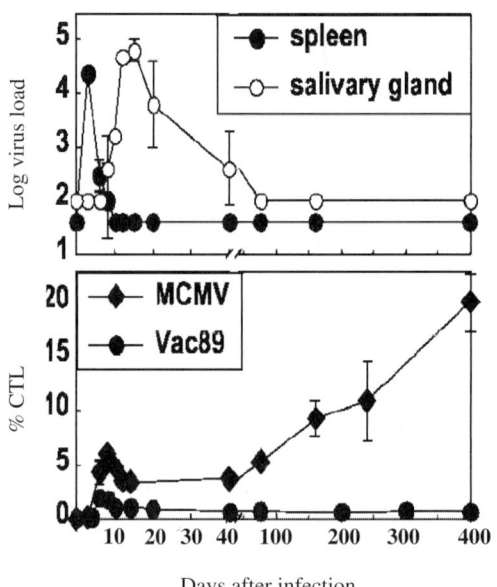

Fig. 5.3. CTL memory inflation in MCMV infection. Data taken from [Karrer et al. (2003)]. The graphs show virus load and CTL responses over time. MCMV virus load was measured in the spleen and salivary gland. pp89 specific CTL responses were measured in the blood. This is a CTL clone that exhibits the inflation dynamics. Mice were infected either with MCMV (diamond symbol), or with a vaccinia virus that expresses the MCMV epitope pp89 (circle symbol). In contrast to MCMV, vaccinia is cleared from the host. Inflation dynamics are only observed in MCMV infected mice, but not in vaccinia virus infected mice. MCMV specific CTL first expand to a peak, subsequently contract, and then inflate.

5.3 An Unusual Pattern of Immunodominance

A different pattern of immunodominance in CTL responses has been observed in murine cytomegalo virus (MCMV) infection. The virus establishes a latent infection and thus persists for the life of the animal [Flemington (2001); Holtappels et al. (2004); Hummel and Abecassis (2002); Reddehase et al. (2002); Scholz et al. (2001)]. The latently infected cells can become reactivated and initiate lytic replication. CTL have been shown to be important for the resolution of acute CMV disease and for keeping the reactivation of latent infection in check [Kaur et al. (1996); Quinnan et al. (1980); Reusser et al. (1999)]. Viral reactivation is initiated from the immediate early (IE) gene complex during latent infection. It is thought that CTL specific for the IE gene products control the infection, because the full replication cycle of the virus is usually not completed. While some CTL responses specific for MCMV showed the typical pattern of CTL dynamics and immunodominance described above, CTL directed against some epitopes show the following unusual dynamics [Karrer et al. (2003); Northfield et al. (2005); Sierro et al. (2005)]. The acute response is characterized by rapid expansion of the CTL, followed by a contraction phase. After that, however, there is a steady accumulation of the specific CTL during the latent phase of the infection (Fig 5.3). As much as 20% of all CD8 T cells can be specific for one CTL epitope one year after infection. These dynamics have been termed "memory inflation" [Karrer et al. (2003)], and could explain why very large numbers of functional CTL specific for human CMV are observed in seropositive individuals long after the resolution of primary infection. Thus, patterns of immunodominance change slowly over time, and a response can become dominant long into the chronic phase of the infection. In the following we discuss these unusual dynamics in the context of mathematical models. We argue that the phenomenon of memory inflation can also be explained by competition for antigenic stimulation. In this case, it is, however, not different CTL clones that compete. Instead, it might be a CTL clone that competes with natural killer cell responses. Natural killer (NK) cells are part of the innate immune system. While they cannot recognize specific viral epitopes, they recognize signs that indicate that a cell is infected and kill the cell [Arase and Lanier (2002); Carayannopoulos and Yokoyama (2004); Dokun et al. (2001); French and Yokoyama (2003)]. NK cells can also proliferate in response to the presence of an infectious agent.

To illustrate this point, we construct a mathematical model for CMV infection. It is based on simple virus dynamics equations and contains the following additional features: (i) the distinction between infected cells that express early viral gene products, and infected cells that express later viral gene products; (ii) a compartment of latently infected cells. The model thus contains the following five variables: susceptible host cells x, infected cells that express early viral gene products y_0, infected cells that express later viral gene products y_1, latently infected cells L, and free virus particles v. It is explained schematically in Fig 5.4, and is given by the following set of differential equations that describe the development of these populations over time.

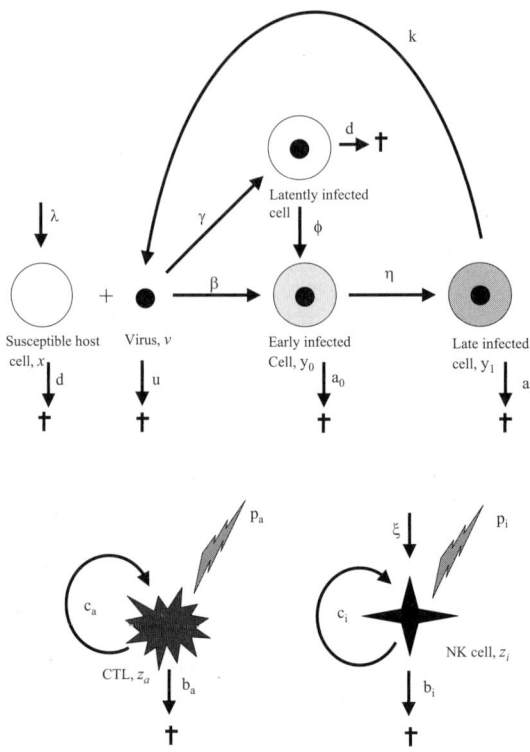

Fig. 5.4. Schematic diagram depicting the basic assumptions that underlie the mathematical model of MCMV infection. The basic model consists of five variables. Susceptible host cells, early infected cells, late infected cells, latently infected cells, and free virus. The details of the model are explained in the text. When a virus particle meets a susceptible host cell and infection occurs, either a productively infected cell or a latently infected cell is generated. The productively infected cells are subdivided into two populations: those which express early gene products (early infected cells) and those which express late gene products (late infected cells). Late infected cells produce new free virus particles and the replication cycle is completed. Latently infected cells are silent but become activated with a certain probability. This gives rise to productively infected cells. In addition to these basic dynamics, the model also contains two types of immune responses. A specific CTL response expands upon antigenic stimulation and exerts antiviral effector activity (through lysis). The second immune response in the model are NK cells. A certain number of NK cells with immediate antiviral activity are always around, and they can also expand in response to antigenic stimulation.

$$\dot{x} = \lambda - dx - \beta xv - \gamma xv, \tag{5.5}$$
$$\dot{y}_0 = \beta xv - a_0 y_0 - \eta y_0 + \phi L - p_a y_0 z_a - p_i y_0 z_i, \tag{5.6}$$
$$\dot{y}_1 = \eta y_0 - a_1 y_1 - p_a y_1 z_a - p_i y_1 z_i, \tag{5.7}$$
$$\dot{L} = \gamma xv - \phi L - dL, \tag{5.8}$$
$$\dot{v} = ky_1 - uv, \tag{5.9}$$
$$\dot{z}_a = \alpha m_n + c_a(y_0 + y_1)z_a - b_a z_a, \tag{5.10}$$
$$\dot{z}_i = \xi + c_i(y_0 + y_1)z_i - b_i z_i. \tag{5.11}$$

Susceptible host cells are produced with a rate λ and die with a rate d. Virus infection results in the generation of a productively infected cell with a rate β, and in the generation of a latently infected cell with a rate γ. Productively infected cells first express early viral gene products (cells denoted by y_0). They die at a rate a_0, and start expressing late gene products with a rate η (cells denoted by y_1). Infected cells that express late viral gene products die with a rate a_1 and produce new virus particles with a rate k. New virus particles decay with a rate u. Latently infected cells die with a rate d, and become activated with a rate ϕ. Activation is assumed to give rise to an infected cell that expresses early viral gene products y_1.

The CTL response is denoted by z_a and is modeled according to the simplest assumptions where antigenic stimulation induces CTL proliferation, and CTL kill infected cells. The variable z_a thus represents the population of effector CTL. It is assumed that CTL can recognize antigen on both types of productively infected cells, i.e. those that express early viral gene products y_0 and those that express late viral gene products y_1. Antigen is not assumed to be recognized on latently infected cells. The CTL response model is given in two parts. In the acute phase it is assumed that naive CTL become stimulated and undergo programmed expansion that involves eight cell divisions, as described by equations (2.18–2.23). Programmed proliferation gives rise to the population of CTL that are denoted by m_n, and these differentiate into effectors z_a with a rate α. Once effectors have been generated, the CTL dynamics are modeled by a simple predator–prey type equation: Upon antigenic stimulation (from both types of productively infected cells), CTL proliferate with a rate c_a. In the absence of antigenic stimulation, they die with a rate b_a. They kill both types of productively infected cells (y_0 and y_1) with a rate p_a. The reason to use predator–prey equations in the post acute phase of the infection is that in the context of virus persistence, the CTL are likely to cycle between the effector and effector memory phenotypes. This should be accurately described by the simple equations (see Chapter 2).

The population of NK cells is denoted by z_i (i stands for innate). It is assumed that reactive NK cells are present independent of the virus. They are produced with a constant rate ξ and die with a rate b_i. Thus, in the absence of infection, NK cells that can react against CMV are present at a level of ξ/b_i. In addition, it is assumed that the population of NK cells can undergo clonal expansion upon antigenic stimulation [Dokun et al. (2001)] with a rate c_i. NK cells kill infected cells with a rate p_i. It is assumed that antigen is recognized on infected cells that express early viral gene products, and on infected cells that express late viral gene products. Latently infected cells are not assumed to be recognized.

In this model, there is a degree of competition for antigenic stimulation between CTL and NK cells. This is because one branch of immunity can suppress virus load and compromise the other. This competition is, however, asymmetric. The development of CTL effector activity depends on antigenic stimulation. The NK cell response can significantly suppress virus load and thereby compromise the development of CTL effector activity. CTL can also suppress virus load. But the generation of NK cell effector activity is not compromised significantly by CTL. This is because a population of reactive NK cells exists even before the infection, and immediate effector activity does not rely on antigen-induced expansion of the NK cell population. Thus, in the following we will examine how the CTL dynamics depends on NK cell activity in the model.

We vary the strength of the NK cell response from high to low. The strength of the NK cell response is captured by the number of reactive NK cells that preexist before the infection (given by ξ/b_i), their effector activity p_i, and their proliferation rate c_i. As mentioned above, if there are a sufficient number of reactive NK cells that preexist before the start of the infection, and if they kill infected cells at a sufficiently high rate, then the basic reproductive ratio of the virus is less than one and an infection is not established. Consequently, a CTL response is not established either. Now assume that the NK cell response cannot prevent the establishment of infection. Now we observe an initial growth phase of the virus population. The extent of this growth depends on how effective the preexisting NK cells are at removing infected cells. The NK cell population will also expand in response to antigenic stimulation. In addition, the population of CMV specific CTL will become stimulated and expand. In the following, we examine how the dynamics of the CTL response depend on the strength of the NK cell response. We will concentrate on the rate of NK cell proliferation (or NK cell responsiveness c_i as a measure of NK cell efficacy. We distinguish between three basic types of CTL dynamics.

(i) The NK cell responsiveness is relatively high and lies above a threshold (Fig 5.5a). We observe an initial expansion phase of the CTL, followed by a decline. This decline will eventually result in the extinction of the CTL population, despite an ongoing persistent infection. The reason is that the NK cell response suppresses virus load to levels that are too low to maintain the CTL response. Whether the specific CTL will go extinct in practice depends on the life span of memory CTL. Experiments suggest that memory CTL can be maintained for long periods of time in

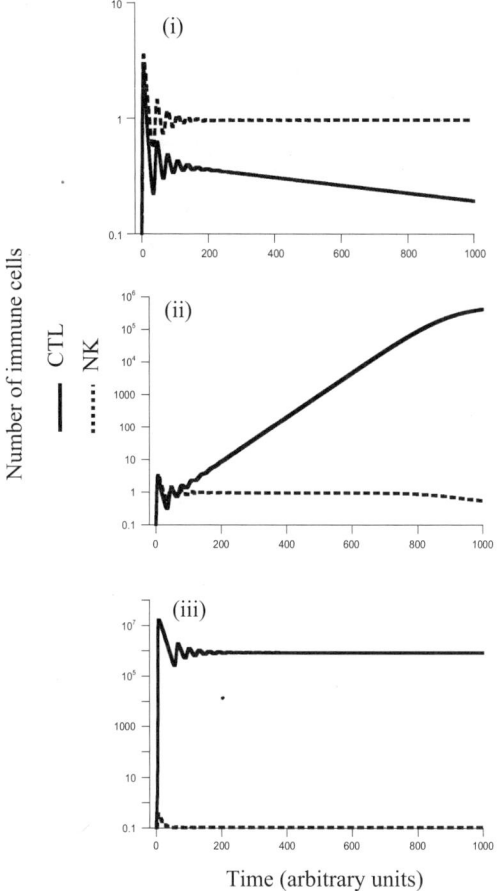

Fig. 5.5. Competition dynamics between CTL and NK cells, based on equations (5.5–5.11). The competition dynamics depend on the relative strength of the CTL and NK cell responses. (i) If the CTL are weak relative to the NK cell response, the NK cells control the virus throughout the course of infection, and resting memory CTL settle around a stable setpoint. (ii) If the CTL is stronger relative to the NK cells, we observe inflation dynamics. At first NK cells control the virus, until the expanding CTL population takes over. (iii) If the NK cells are too weak relative to the CTL, the CTL immediately control the infection and no inflation is observed. Parameters were chosen as follows: $\lambda = 10$, $d = 0.1$, $\beta = 1$, $a_0 = 0.1$, $a_1 = 0.2$, $p_0 = 1$, $p_1 = 0.000001$, $k = 1$, $u = 1$, $c_1 = 15.5$, $b = 0.1$, $\gamma = 0.5$, $\phi = 0.1$, $\eta = 0.01$, $\xi = 0.01$, $r = 1$, $\delta = 10$. For (i) $c_0 = 1$, (ii) $c_0 = 12$, (iii) $c_0 = 14$.

the absence of antigenic stimulation. In this case, the CTL would not go extinct in a realistic period of time, but persist as a population of resting memory CTL that decline at a very slow rate (Fig 5.5a).

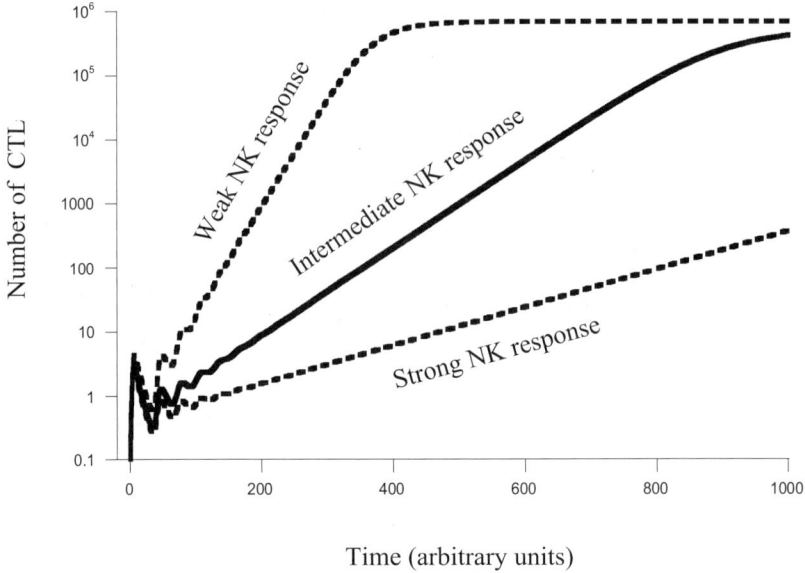

Fig. 5.6. Degree of CTL inflation depending on the protectiveness of the NK cell response. The weaker and less protective the NK cell response, the stronger the degree of CTL inflation that is observed. The reasons in as follows. The weaker the NK cells, the less virus load is reduced after acute infection. The less virus load is reduced, the higher the amount of antigenic stimulation for the CTL during chronic infection, and the faster the rate of CTL expansion. As mentioned in Fig 5.4, if the protectiveness of NK cells falls below a threshold and is too low, such that CTL are more effective at reducing virus load already during acute infection, then CTL inflation does not occur and these considerations do not hold. Simulations are based on equations (5.5–5.11). Parameters were chosen as follows: $\lambda = 10$, $d = 0.1$, $\beta = 1$, $a_0 = 0.1$, $a_1 = 0.2$, $p_0 = 1$, $p_1 = 0.000001$, $k = 1$, $u = 1$, $c_1 = 15.5$, $b = 0.1$, $\gamma = 0.5$, $\phi = 0.1$, $\eta = 0.01$, $\xi = 0.01$, $r = 1$, $\delta = 10$, $c_0 = 10, 12 \& 13$.

(ii) The NK cell response is weaker, and antigenic stimulation can maintain the CTL in the long-term (Fig 5.5b). We again observe an initial expansion of the CTL population, followed by a contraction phase. But now, the contraction phase is followed by a steady increase of the number of specific CTL during the chronic phase of the infection (Fig 5.5b). This may correspond to the CTL inflation dynamics. The reason is as follows. The NK cell response is initially more effective than the CTL because it can achieve higher levels of lysis earlier on. Thus, NK cells play the dominant role in this acute phase and downregulate the virus population. This prevents the CTL from expanding fully. Consequently, they contract in the face of limited antigenic stimulation. The dynamics between the virus population and the NK cells oscillates to a steady state. Virus load at this steady state is high enough so that the CTL become stimulated and expand during the chronic phase. CTL expansion continues until the number of CTL is high enough such that CTL-mediated killing becomes the dominant immune effector mechanism. Then the CTL settle around a steady state and control the virus population (Fig 5.5b). The NK cell response is expected to decline to a certain extent at this stage. The inflation dynamics are determined by the following factors. The strength of NK cell mediated virus control can determine the degree of inflation (Fig 5.6). The weaker the NK cell response, the faster the rate of inflation. That is, the CTL population expands relatively slowly if the NK cell population is stronger, and faster if the NK cell response is weaker. The weaker the NK cell response, the higher the antigenic drive during chronic infection, and this allows for more pronounced CTL expansion (Fig 5.6). Another factor that influences the amount of inflation can be the rate of CTL mediated effector activity. CTL inflation is more pronounced if the rate of CTL-mediated activity is weaker. This is because with weaker CTL-mediated activity, a higher number of specific CTL are required to achieve CTL-mediated control of the virus population.

(iii) Finally, if the NK cell response is less effective at reducing acute virus load compared to the CTL response, we do not observe CTL inflation dynamics (Fig 5.5c). In this case, CTL settle around a stable memory level after the acute phase of the infection. The difference to the first scenario described above is that now the memory cells are expected to be activated and not resting. The reason is that NK cells now do not play a significant role in limiting acute virus growth. Consequently, the CTL are immediately the dominant immune response that drive the dynamics during acute infection and suppress virus load.

In summary, the dynamical interplay between NK cells, CTL, and the virus population can lead to the phenomenon of CTL inflation where patterns of immunodominance change slowly over the long-term during the chronic phase of infection. If NK cells are more efficient than CTL at reducing virus load during acute infection but fail to keep virus load at sufficiently low levels during chronic infection, then we expect that CTL first expand to a limited peak, contract, and then inflate. A major factor that determines the extent of CTL inflation is the strength of the NK cell mediated control of chronic virus load. If the NK cells response is strong, it keeps the virus at relatively low levels during the chronic phase of the infection. This only provides a limited antigenic stimulus for the expansion of CTL, and the extent of CTL inflation is low. A weaker NK cell response allows higher virus loads during the chronic phase of the infection. This allows for a higher antigenic stimulus for the CTL. Consequently, CTL inflation is more pronounced. In addition, the weaker the rate of CTL-mediated effector activity, the larger the degree of inflation because more specific CTL are required to control the virus population. The prediction that the extent of NK cell mediated protection can shape the dynamics of CTL memory inflation has recently confirmed by experimental data from MCMV infected mice (Wodarz et al, in preparation).

While the model has only considered a single CTL response, multiple CTL clones directed against different epitopes are observed *in vivo*. Depending on the CTL responsiveness and the rate of CTL-mediated effector activity, some CTL responses may inflate while others do not. For example, the amount of virus maintained by the NK cell response may be too little to achieve expansion for some CTL clones during chronic infection, while it will be sufficient for others. Consequently some clones might remain as resting memory cells at a stable level, while others inflate, further complicating the patterns of immunodominance. This has been observed in experimental data from MCMV infected mice [Karrer et al. (2003)].

5.4 Summary

In this chapter we reviewed mathematical models that have aimed to explain patterns of immunodominance observed *in vivo*. Of central importance is the concept of competition between immune responses. CTL clones directed against different epitopes of the same virus compete for antigenic stimulation. The more efficient a competitor, the lower the level to which it can reduce virus load, and the more it can suppress the expansion of competing CTL clones. This is how the dominance of one CTL clone can come about. The dominance of a single CTL clones is expected to be observed especially in the context of a weak response. If the response is stronger, the degree of competition between the CTL clones is reduced, and we expect to see a broader CTL response where several CTL clones coexist over a prolonged period of time. Antigenic heterogeneity can also contribute to a broadening of the CTL response. Finally, we examined an unusual pattern of immunodominance where certain CTL clones can be present initially at low levels and then slowly increase in abundance during the chronic phase of the infection. Such inflation dynamics have been observed in the context of murine MCMV infection, and can be explained by the competition between CTL responses and NK cell responses for antigenic stimulation.

6
Multiple Infections and CTL Dynamics

When CTL dynamics are analyzed, we usually consider a CTL response in the context of a single specific virus. Upon infection, the CTL response expands, attains effector activity, fights the virus, and differentiates into memory cells. Memory cells survive at elevated levels for a prolonged period of time after the resolution of infection. However, hosts are exposed to a wide variety of infections over their lifespan. Each infection can potentially elicit CTL responses that expand and build memory. This brings up a problem. As the host gets infected by different pathogens, the total number of CTL would increase over time as more and more populations of memory cells are created. However, this does not occur. Instead, experiments have shown that a given infection can result in the decline of the CTL memory population that was established in response to a previous unrelated infection (Fig 6.1) [Liu et al. (2003); Selin et al. (1999); Selin et al. (1996); Welsh et al. (1995)]. Thus, already established CTL memory is diminished upon exposure to heterologous antigen. It appears that the second infection activates the CTL memory population that was established in response to a previous infection trough bystander effects. Once activated, however, these CTL do not receive any further survival signals because they are not specific for the antigen that is currently present in the system. Consequently, they die and the population of memory cells declines.

These findings show that CTL responses to individual viruses do not occur in isolation, but influence each other. If long-term persistence of memory CTL would primarily serve to protect the host against secondary viral challenges, then the only implications of these findings would be that protection against rechallenge diminishes over time while the host experiences infection with heterologous viruses. On the other hand, if long-term persistence of memory CTL is required to achieve virus clearance or long-term control of the infection (see Chapter 2), then the finding that exposure to heterologous viruses can lead to the attrition of previously established CTL memory has important implications for the ability of the immune system to deal with multiple infections at the same time. This chapter reviews mathematical models that have addressed this issue.

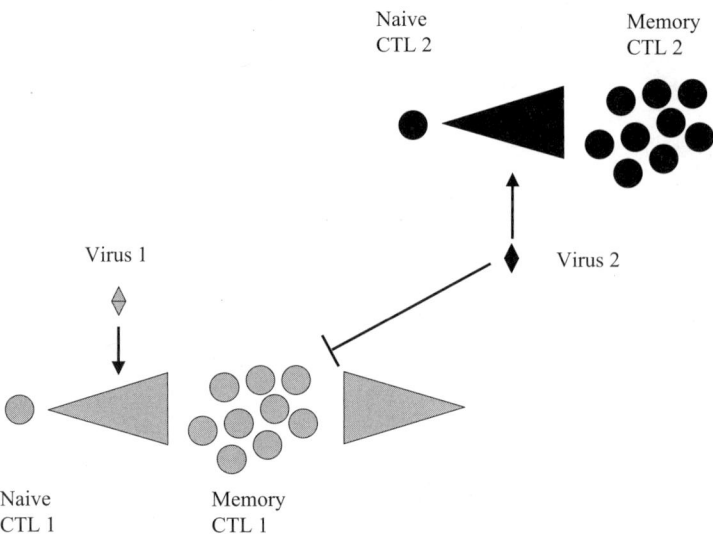

Fig. 6.1. Schematic illustration of memory attrition by heterologous antigen. If the host is exposed to a given infection, it mounts a CTL response and builds memory. If the host is subsequently exposed to another infection, it mounts a CTL response and builds memory against this second infection. In addition, the second virus induces a decline of the memory CTL against the first virus. This is the central concept of the current chapter.

6.1 Mathematical Model

We construct a simple model taking into account two basic variables. The virus populations, and the memory CTL populations specific for the respective viruses. We assume that the host can be infected by n different viruses during its life time. Virus of type i is denoted by v_i, and the memory CTL response specific to that virus is denoted by z_i, where $i = 1..n$. The model is given by the following set of ordinary differential equations.

$$\dot{v}_i = r_i v_i \left(1 - \frac{v_i}{k_i}\right) - p_i v_i z_i, \tag{6.1}$$

$$\dot{z}_i = c_i v_i z_i - b_i z_i - z_i \sum_{\substack{j=1..n \\ j \neq i}} \alpha_j v_j. \tag{6.2}$$

This model uses a simplified equation to describe virus growth [Wodarz (2001a)]. Instead of taking into account explicitly the number of infected and uninfected cells, as well as the population of free viruses, we capture the growth of the virus population in a single equation (2.1). The virus replicates at a rate r_i. Replication is density dependent, limited by the carrying capacity

k_i. The carrying capacity corresponds to target cell limitation. Virus replication is inhibited by the specific CTL response at a rate p_i. The memory CTL response becomes activated and proliferates in response to its specific antigen at a rate c_i. In the absence of antigen, the CTL response has an average life span of $1/b_i$. In addition, the model assumes that heterologous antigenic stimuli v_j reduce the memory CTL response to virus i (z_i) at a rate α_j.

Note that the model makes a number of assumptions that are worth mentioning explicitly: (i) The n different viruses are assumed not to be in competition with each other. That is, every virus has its own set of target cells without significant levels of direct interference. This is a reasonable assumption if the viruses are distinct pathogens. If the antigenically heterologous virus only is a different strain of the same pathogen this assumption is valid if virus load remains low and competition is not the driving force underlying the dynamics (see discussion). (ii) For analytical simplicity it is assumed that the impairment by a viral strain is the same for all strains, i.e. the α_j values are the same for all i. (iii) In the model, the CTL response to a given infection is impaired directly by the presence of antigenically heterologous viral stimuli. An alternative mechanism could be that immune responses have an effect on each other. This would be similar to Jerne's network hypothesis [Hoffman (1975); Jerne (1974a); Jerne (1974b); Urbain (1986)] according to which immune cell type 1 can act on cell type 2 and vice versa. This can be viewed as a reciprocal predator–prey interaction in which a given immune cell type is both predator and prey to another, and vice versa. There has been considerable debate and controversy regarding the network hypothesis, and the applicability of such a regulatory mechanism could be different in the resting and the active immune state [Anderson and May (1991)]. Here the network hypothesis is not further pursued, since mathematical modeling approaches [Anderson and May (1991)] suggest that it might not be able to fully account for the observed experimental results.

6.2 Virus Control and Antigenic Heterogeneity

We start by summarizing the properties of CTL responses specific for a single virus population in isolation (For now we omit subscripts for simplicity). If $r > 0$, then the virus population grows during primary infection. If $ck > b$, virus growth is followed by expansion of the CTL response. The rising CTL response reduces virus load that eventually settles at an equilibrium level described by $v^* = b/c$. The level of CTL at equilibrium is given by $z^* = r(1 - v^*/k)/p$. According to the model, virus load at equilibrium is influenced by two immunological parameters. (i) A low virus load is promoted by a high CTL responsiveness c. (ii) Low virus load and long-term control of an infection requires a long life span of the CTL response in the absence of antigen, i.e. a high value of $1/b$. Hence long-term virus control or clearance requires antigen-independent persistence of CTL memory. This is important because antigen-independent persistence of memory CTL ensures that immunological pressure is maintained even if virus load declines to low levels. These notions are the same as those described in Chapter 3.

These considerations apply as long as a given virus population is cleared before the host is infected with an antigenically heterologous virus. However the situation becomes more complicated if the host is faced with more than one virus infection at the same time. There might be multiple acute infections simultaneously present, or the host might harbor a collection of persistent infections that can be immunologically controlled. In such a setting, a given virus population can influence immune responses to the other pathogens. If the CTL response has to deal with more than one infection at the same time, the characteristics of the response required to result in virus control or clearance can be altered. Equilibrium virus load for a given infection v_i in the presence of the specific CTL response z_i is described by

$$v_i^* = (1/c_i)\left(b_i + \sum_{\substack{j \neq i}}^{j=1..n} \alpha_i v_j^*\right).$$

Thus, the abundance of virus i also depends on the collective abundance of the other viruses in the host. More importantly, if

$$\sum_{\substack{j \neq i}}^{j=1..n} \alpha_i v_j^* > b_i,$$

then a long life span of memory CTL in the absence of antigen (low value of b_i) loses its ability to contribute to virus control. In this case, reduction of virus load is only promoted by a high CTL responsiveness c_i. Thus, although the CTL memory response still has in principle the capacity to persist in the absence of antigen (low b), interference by the heterologous antigenic stimuli renders this memory ineffective at maintaining strong immunological pressure at low virus loads. At low loads, the degree of interference from heterologous stimuli is stronger than the amount of specific antigenic stimulation, resulting in reduction of CTL-mediated pressure. Hence, overall immunity is compromised, and the chances of virus eradication, as well as the level of control, is reduced.

6.3 Two Heterologous Infections

Here we assume that a given virus population v_1 has established persistent infection and is efficiently controlled by a CTL memory response z_1. Failure to clear the infection could have a variety of reasons. Examples are infection of sites that are difficult to access by the immune response, or viral latency. We investigate the consequence of infection with a heterologous virus v_2 for the

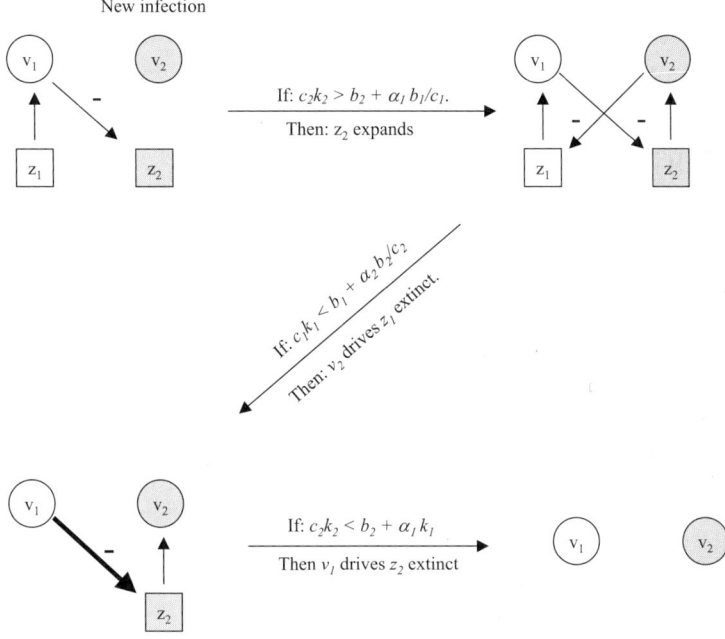

Fig. 6.2. Dynamics of two antigenically heterologous infections. Assume that virus population v_1 is controlled by the CTL response z_1. The schematic diagram demonstrates possible outcomes when adding a second virus infection v_2. Either both infections are controlled by their respective CTL memory responses, or one or both responses are driven extinct, resulting in uncontrolled virus growth. For detailed explanation, see text.

degree of antiviral immunity to both infections (Fig 6.2). The CTL response against the heterologous virus z_2 can expand if $c_2 k_2 > b_2 + \alpha_1 b_1 / c_1$. If this condition is fulfilled and the CTL response z_2 does invade, then the outcome of infection is as follows. If $c_1 k_1 > b_1 + \alpha_2 b_2 / c_2$ then both infections are controlled by their respective CTL responses. In the opposite case, the CTL response z_1 goes extinct, and virus population v_1 reaches its carrying capacity k_1. Since increased load of virus population v_1 has an increased adverse effect on the CTL response z_2, immunological control of virus population v_2 is also reduced. If $c_2 k_2 > b_2 + \alpha_1 k_1$ then the CTL response z_2 is still maintained and can control the virus population v_2 to a certain degree. In the opposite case, the CTL response z_2 also goes extinct, and the virus population v_2 reaches its carrying capacity k_2.

Similarly, if the CTL response z_2 cannot invade in the first place, virus population v_2 will replicate up to its carrying capacity k_2 and reduce immunity to virus population v_1. If $c_1 k_1 > b_1 + \alpha_2 k_2$, then the CTL response z_1 is maintained, otherwise it goes extinct and $v_1 = k_1$.

This simple analysis demonstrates how the invasion of a heterologous pathogen can upset the ability of CTL to control infections. It can potentially lead to the collapse of the CTL response against both infections.

6.4 Multiple Heterologous Infections

Here we extend the above analysis to account for multiple persistent infections. We investigate how accumulation of persistent infections influences overall immunity and virus control, and discuss whether there is a limit to the number of infections the immune system can deal with (Fig 6.3). We assume that there are a number of n persistent virus infections v_i that can infect the host over time and that can be controlled by their respective specific CTL responses z_i. We start by considering the simplest case assuming that viral and host parameters are the same for each infection and the respective CTL response (i.e. $r_i = r; k_i = k; p_i = p; c_i = c; b_i = b; \alpha_i = \alpha$). Fig 6.3 shows how total virus load and the total number of memory CTL depends on the number of infections present in the host. Increasing the number of infections n results both in an increase in virus load and the total number of memory CTL. For

$$n > (1 - c/\alpha) - \frac{\sqrt{kb(\alpha + c)}}{k\alpha},$$

the total number of memory CTL declines with an increasing number of infections. For

$$n > (1 + c/\alpha) - \frac{b(c + \alpha)}{ck\alpha},$$

CTL memory collapses and the immune system loses control of all virus infections. It is interesting to consider the rate of increase in total virus load when infections are accumulated (when n is increased, Fig 6.3). For $n < (1/2)(1 + c/\alpha)$ the increase in total virus load with the addition of new infections is less than exponential (Fig 6.3). However, this trend is reversed if $n > (1/2)(1 + c/\alpha)$: if this threshold is crossed, then the increase in total virus load becomes greater than exponential when new viruses are added to the system (i.e. when n is increased, Fig 6.3). Hence, if the number of infections crosses this threshold, CTL memory starts to lose the ability to keep the viruses in check and this culminates in a decrease of the CTL memory population and eventually in extinction of CTL memory.

These basic patterns also underlie the more complicated and realistic case assuming that host and viral parameters differ between individual infections (Fig 6.4). An important parameter in this respect is the strength of the memory CTL responses. In the model this is described by $c_i k_i$, i.e. it is a combination of the rate of CTL activation and the level of antigenic stimulation provided by the virus. For the purpose of analysis, we rank the CTL specific for the different viruses according to their rate of expansion, so that

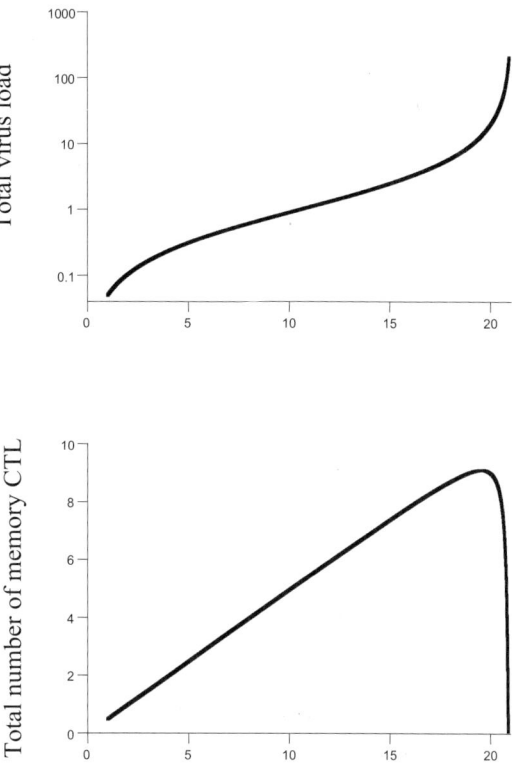

Fig. 6.3. Effect of the number of infections n on total virus load and the total number of memory CTL in the neutral model, assuming that parameters are identical for each infection. An increase in n results in an increase in virus load. This increase is less than exponential if n lies below a threshold defined in the text. If n lies above that threshold, the increase in virus load is faster than exponential. An increase in n also results in an increase in the total number of memory CTL. If n crosses a threshold, the overall number of memory CTL decreases and finally goes extinct. For mathematical details, see text. Simulations are based on equations (6.1–6.2). Parameter values were chosen as follows: $r_i = 0.5$, $p_i = 1$, $k_i = 10$, $b_i = 0.1$, $c_i = 2$, $\alpha_i = 0.1$.

$c_1 k_1 > c_2 k_2 > c_3 k_3 > \ > c_n k_n$. We can then test each successive CTL response z_i for its ability to persist in the face of the other virus infections. The CTL response z_i directed against virus infection v_i can persist and control the infection if

$$\sum_{\substack{j=1..n \\ j/neqi}} \alpha_j v_j^* < c_i k_i - b_i.$$

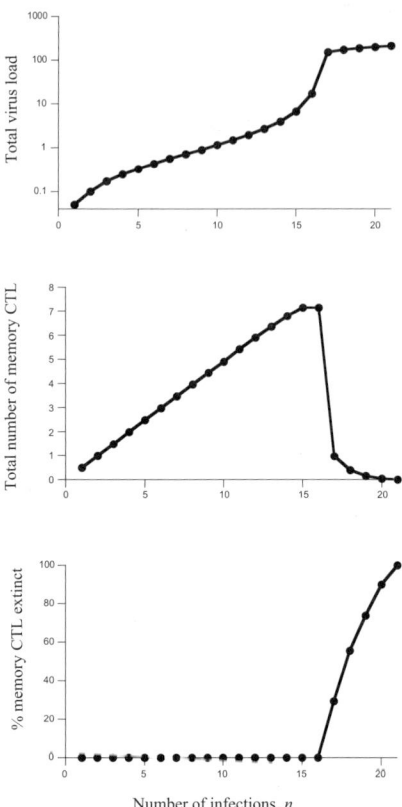

Fig. 6.4. Effect of the number of infections n on total virus load and the total number of memory CTL in the realistic model, assuming that parameters differ between the infections. The general pattern is the same as for the neutral model shown in Fig 6.3. The main difference is that in this more realistic case, not all CTL memory responses necessarily go extinct when the number of infections n crosses a certain threshold. If n crosses a threshold, only a fraction of the CTL memory responses are driven extinct. With each additional virus infection added to the system, a higher fraction of the CTL memory responses goes extinct, until all viruses can grow uncontrolled. The simulation assumes that the main difference between the infections is the CTL responsiveness c_i. Values were assigned according to $c_i = 2 - 0.05i$, where $i = 1..n$. Other parameter values were chosen as follows: $r_i = 0.5$, $p_i = 1$, $k_i = 10$, $b_i = 0.1$, $\alpha_i = 0.1$. Simulations are based on equations (6.1–6.2).

Thus, establishment of the memory CTL response z_i is promoted by a low total virus load of heterologous pathogens, as well as by a small negative effect of these heterologous infections on the CTL response z_i. If the above condition is not fulfilled, then the CTL memory response z_i is not established, and the virus population v_i reaches its carrying capacity k_i. This in turn significantly increases overall virus load, and this can have an increased negative effect on all other memory CTL responses present. If overall virus load is high enough, this can lead to a chain reaction resulting in extinction of further CTL responses (Fig 6.4). In the worst case, this chain reaction culminates in extinction of the entire CTL memory population and uncontrolled replication of all viruses present (Fig 6.4). This would be equivalent to death of the host.

6.5 Experimental Studies

In a set of experiments, mice were first infected and primed with influenza A virus [Liu et al. (2003)]. That is, they generated a CTL response and CTL memory that persisted in the long-term. Subsequently, the mice were infected with gamma herpes virus $\gamma HV68$. Mice were sampled at 35, 60, and 100 days after gamma herpes virus infection. At every time point tested, the influenza specific memory CTL frequencies were significantly reduced for a number of tissue sites in the mice. This effect was selective for the influenza specific CTL memory population. Estimates of the total CTL count showed that there was no significant reduction as a result of gamma herpes virus infection. It was also checked whether the presence of the large, influenza specific memory CTL population at the time of gamma herpes virus infection would modify the new gamma herpes virus specific response. No significant effect was determined. Therefore, the negative influence of one response on the other is asymmetric. A new response can diminish a previously established memory CTL population, but the previously established memory CTL population does not have a negative impact on a new response.

This experiment was specifically simulated using the mathematical model described above (6.1–6.2). While gamma herpes virus does does establish latent infection, this has not been included in the model for simplicity because it does not influence the results in question. The simulation (Fig 6.5) starts with one infection that is resolved by a specific CTL response. This corresponds to influenza virus in the experiments. After the virus is cleared, the memory CTL are long-lived and only decay at a very slow rate. During this memory phase, a second virus (gamma herpes virus) is added to the system. The maximum virus load of the lytic replication is resolved by the CTL, leading to long-term memory and control. During the time that the gamma herpes virus load increases, the CTL memory response to influenza virus is reduced (Fig 6.5). As the gamma herpes virus load is diminished, the CTL memory response against influenza virus is no longer negatively affected. The experimental data showed that the CTL response against gamma herpes virus is not selectively impaired, despite the presence of high initial numbers of influenza virus specific CTL. This is in accord with the model simulation, assuming that the heterologous antigen itself, and not CTL competition, is the reason for the reduction of CTL memory.

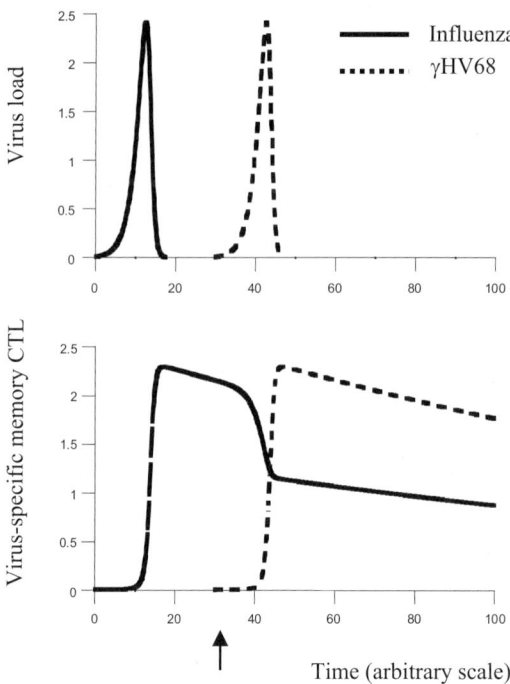

Fig. 6.5. Simulation of the experiments in which mice were first primed with influenza virus and then challenged with $\gamma HV68$ (gamma herpes virus). Simulations are base on equations (6.1–6.2). The arrow indicates the time of $\gamma HV68$ infection. A new CTL response against $\gamma HV68$ develops, and the previous CTL memory that was induced by influnza, declines. Parameter values were chosen as follows: $r_i = 0.5$, $p_i = 1$, $k_i = 10$, $b_i = 0.005$, $c_i = 0.5$, $\alpha_i = 0.05$. For simplicity, the parameters were assumed to be the same for both infections.

6.6 Coinfection: Viruses and Bacteria

It is often the case that viral infections, especially with respiratory viruses, are accompanied by bacterial coinfections. According to the model, coinfection can reduce overall immunity, and this can result in delayed clearance, or in failure to clear either infection (Fig 6.6). This weakened immunity in turn opens up the possibility for other viruses to invade the host and to deteriorate the situation. Hence, especially in older patients, or in patients with a compromised immune system, it could be helpful to administer antibiotics even if symptoms are caused by a viral infection. Protection from bacterial coinfections ensures that the CTL response can work with highest efficacy to resolve the viral disease. If there already is a bacterial coinfection present in the host, inhibition of bacterial growth reduces exposure of the immune system to bacterial antigen, and this might enhance the CTL memory response against the virus (Fig 6.6). Hence, inhibition of bacterial growth could result in more efficient antiviral immunity and in faster resolution of the viral disease.

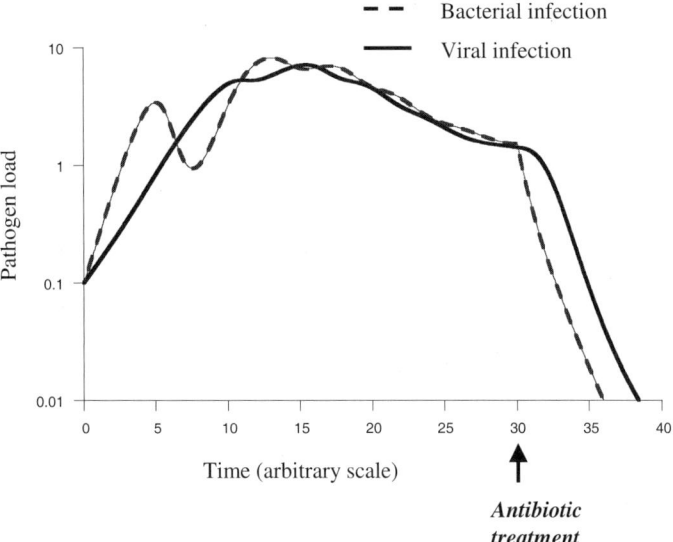

Fig. 6.6. Coinfection by a virus (v_1) and a bacterial pathogen (v_2). The viral and bacterial infections are described by the same equations (6.1–6.2). If the amount of CTL memory attrition is significant, coinfection can result in impaired or delayed resolution of both pathogens. Treatment of the bacterial infection (e.g. with antibiotics, arrow) reduces the amount of bacterial antigen displayed to the immune system. This strengthens the CTL memory response to the viral infection that can now be driven extinct. Treatment of the bacterial infection was modeled by setting $r_2 = 0$, and introducing a term describing removal of the bacteria, $\gamma = 0.5$. Other parameter values were chosen as follows: $r_1 = 1$, $r_2 = 0.5$, $c_1 = 0.5$, $c_2 = 0.6$, $k_i = 10$, $p_i = 1$, $b_i = 0.1$, $a_i = 0.5$.

6.7 Vaccination

The fact that an increase in the number of antigenic stimuli experienced by the immune system at the same time can result in overall compromised immunity has obvious implications for vaccination strategies. Immunization is the most effective way of preventing infectious disease. Traditionally, there are two main types of vaccines: live virus vaccines and inactivated virus vaccines. Live avirulent vaccines have been most successful and have significantly reduced the incidence of several important diseases. The reason for the success of these vaccines is that the live attenuated virus replicates in the recipient. However, this can also have negative effects on overall immunity, since from an immunological point of view this antigenic exposure is equivalent to a natural infection. Even inactivated virus vaccines could in principle have negative effects on overall antiviral immunity since application of multiple high doses of the antigen could lead to attrition of the memory CTL population against other infections. According to the arguments presented here, it is not a good strategy to vaccinate people against as many infections as possible. Vaccination should be administered if the benefit to the individual achieved by the immunization outweighs the cost to overall immunity of that individual (due to a possible reduction in previously established CTL memory populations). Hence, patients currently dealing with an acute infection should abstain from vaccination, since the additional exposure to heterologous antigen could reduce the efficacy of CTL memory to resolve this acute infection. Similarly, patients harboring pathogenic persistent infections such as HIV, HTLV, or hepatitis should carefully consider possible immunizations.

From a broader perspective we could be faced with a potential dilemma when thinking about vaccination strategies: While vaccination could be detrimental for individuals who have already experienced many different antigenic stimuli, vaccinating against as many infections as possible could be beneficial for the society as a whole, since it would effectively stop the spread of the pathogen within the population. This problem merits further investigation by mathematical models.

6.8 The Immune Phenome and Aging

The mathematical analysis of multiple infections in overall CTL dynamics illustrates how how the life time immune effector and memory phenome will be shaped by the overall antigenic (particularly the infection) history of the individual. The consequences of those in high infection load environments are potentially disastrous. An obvious example are people living in the developing world and those involved in high risk behavior such as intravenous drug use. As pointed out in the last section, such interactive effects need to be kept in mind as we attempt to develop effective vaccination strategies for these vulnerable populations. Furthermore, the model has important implications for antigenically variable pathogens (e.g. HIV or HCV) that expose the immune system to a wide variety of different antigenic stimuli over a relatively short period of time. This could result in the persistence of the virus variants, with overall virus load increasing as more antigenic strains are produced. When the number of antigenic strains crosses a threshold, CTL memory will decline and eventually collapse.

These notions could also account for the observation that the efficacy of antiviral immunity decreases with age. With advanced age, people become more susceptible to viral infections, and recall responses have been observed to be less efficient compared to younger age groups [Effros and Walford (1983); Fagiolo et al. (1993)]. During the life time the host becomes infected with a wide variety of viruses and antigen might persist for prolonged periods of time. Other viruses establish persistent infections. Thus, as the host ages, the number of heterologous antigenic stimuli presented to the immune system could increase, resulting in progressive weakening of CTL memory and antiviral immunity. CTL memory responses might also be required for tumor surveillance, and might prevent cancer growth [Xiang et al. (1999)]. In this case, the emergence of clinical tumors could also be the consequence of weakened CTL memory caused by accumulation of too many antigenic stimuli over time.

6.9 Summary

This chapter has stressed that specific virus immune system interactions should not be considered in isolation, but that heterologous infections can activate and diminish previously established CTL memory in a nonspecifc manner through bystander effects. In particular, the models discussed here suggest that the negative effect on previously established memory is mediated directly through the new virus population, and not through competition between the different CTL responses. The model assumptions and predictions have been tested by experiments in which mice were sequentially challenged with influenza virus and gamma herpes virus. These concepts have long reaching consequences for the host. On the simplest level, the host loses memory-mediated protection against rechallenge if memory gets deleted by exposure to heterologous antigen. Moreover, simultaneous exposure to multiple pathogens can render the host less effective at dealing with those pathogens. Impaired memory responses can result in a reduced efficacy of the CTL at controlling or eliminating viruses and pathogens in general. Once the host has been exposed to too many infections, the model suggests that the overall capability of the CTL are reduced, until the CTL populations collapse and fail to fight any infection successfully. This trend might be observed during old age.

7
Control versus CTL-Induced Pathology

A major mechanism by which CTL fight viruses is the lysis or killing of infected cells. This removes the source of virus production and contributes to virus clearance. As discussed in Chapter 8, lysis of infected cells can be essential for the resolution of infection [Jeffery et al. (1999); Kagi et al. (1996); Kagi et al. (1995b); Saah et al. (1998); Schmitz et al. (1999)]. This form of antiviral activity can, however, also have a negative impact on the host [Lehmann-Grube (1971); Moskophidis et al. (1993c); Zinkernagel (1996)]. If many cells are infected, then lysis of these cells can lead to a significant amount of tissue damage. This can result in pathology, and even in the death of the host. Such damage, brought about by CTL responses that fight an infection, is referred to as *CTL-induced pathology*.

CTL-induced pathology is an example of a variety of conditions in which the immune system can harm its own body. It is important to distinguish it from the class of diseases that are known as autoimmune diseases [Fujinami (2001); Matzinger (1998); Rouse and Deshpande (2002); Seewaldt et al. (2000); Zinkernagel (1993)]. In autoimmune diseases, inappropriate immune responses against proteins of the host's own body are triggered, and this results in pathology. It is unclear how inappropriate responses develop. A viral infection can be the reason. For example, the virus can carry an antigen that mimics a host antigen, and this molecular mimicry can induce the disease. With CTL induced pathology, an appropriate CTL response is triggered that recognizes viral antigen and that can contribute to the resolution of the infection. However, in the process of fighting the virus, a large number of tissue cells can become killed if the virus has infected many cells, and this is the reason for pathology.

The concept of CTL-induced pathology is best defined in the experimental LCMV infection of mice [Lehmann-Grube (1971); Moskophidis et al. (1993c); Zinkernagel (1996)]. As reviewed in Chapter 1, LCMV infection is characterized by a variety of outcomes that range from viral clearance to CTL exhaustion [Moskophidis et al. (1995b); Moskophidis and Kioussis (1998); Moskophidis et al. (1993a); Moskophidis et al. (1993c); Zinkernagel (1996);

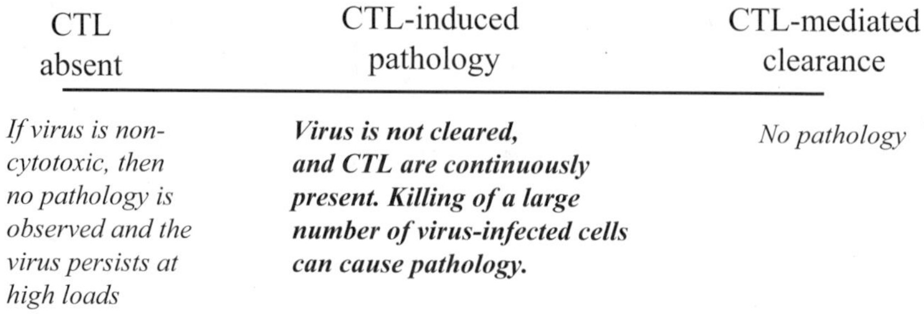

CTL absent	CTL-induced pathology	CTL-mediated clearance
If virus is non-cytotoxic, then no pathology is observed and the virus persists at high loads	***Virus is not cleared, and CTL are continuously present. Killing of a large number of virus-infected cells can cause pathology.***	*No pathology*

Fig. 7.1. Schematic diagram that puts the concept of CTL-induced pathology into context. A variety of outcomes can be observed. On the one hand, the CTL response can be absent and can thus not cause pathology. If the CTL response clears the virus, it also does not induce pathology. If CTL fail to clear an infection and are continuously present and killing infected cells, CTL-induced pathology can be observed.

Zinkernagel et al. (1977)]. The outcome of infection is determined by the initial virus load, the replication rate of the virus, and host parameters such as the strength of the CTL response. If the CTL response is not exhausted, but also fails to control the infection, the outcome is CTL-induced pathology where the mouse loses weight, wastes, and dies (Fig 7.1). The antiviral CTL have been shown to be the reason for this pathology. It is thought that such immunopathology also plays a role in the pathogenesis of human infections, but the evidence is often less clear. For example, it is not understood whether the symptoms of the common cold are caused by the virus itself, or by the CTL that fight the virus and therefore cause tissue damage. Based on the results obtained from LCMV infection, it has been suggested that CTL-induced pathology could contribute to the development of AIDS [Zinkernagel (1994); Zinkernagel (1995); Zinkernagel and Hengartner (1994); Zinkernagel et al. (1999)]. Similarly, CTL-induced pathology could contribute to liver disease in hepatitis B and C virus infections (HBV and HCV) [Guidotti et al. (1994a); Guidotti et al. (1999a); Guidotti and Chisari (1996); Guidotti et al. (1999b)].

This chapter will review how mathematical models have helped to define the conditions under which CTL have a net positive effect for the host, and when CTL-induced pathology occurs. Theoretical predictions will be discussed in the context of experimental data from LCMV infection. Finally, we explore implications for HIV.

7.1 Basic Mathematical Insights

This section describes some basic mathematical results that give insights into the conditions under which CTL-induced pathology occurs, and when CTL

7.1 Basic Mathematical Insights

have an overall beneficial effect on the host. This can be done in the context of the basic equations that describe infection dynamics (model 2.2–2.4). The CTL response is given by equation (2.11) [Wodarz and Krakauer (2000)]; that is, CTL are assumed to expand in response to antigenic stimulation. The principles described here do not, however, depend on this particular mathematical form to describe CTL dynamics. The degree of pathology is measured by the sum of the number of uninfected and infected host cells that are targeted by the virus $x+y$; in other words, the total number of tissue cells. In the absence of infection, all cells are uninfected, and the equilibrium number of tissue cells is given by $x^{(0)} = \lambda/d$. If an infection becomes established, the system converges to a new equilibrium where the total number of tissue cells is less than or equal the number of tissue cells in the absence of infection ($x+y <= x^{(0)}$). The degree of pathology is measured by the fraction of tissue cells that remain at equilibrium in the presence of the infection. We analyze equilibria because CTL-induced pathology occurs mostly in persistent infections. There might be transient pathology in acute infections that are eventually cleared, such as the common cold. The properties of pathology in acute and transient infections are, however, very similar to those derived from the equilibrium analysis.

We start the analysis by assuming that the virus is noncytotoxic (i.e. does not kill the infected cell, $a = d$). An example of this is LCMV infection of mice [Lehmann-Grube (1971)]. Two parameters are important for determining the degree of CTL induced pathology. These are the efficacy of the CTL response (described by the immune resonsiveness c and the rate of CTL-mediated lysis p) and the replication rate of the virus (described by the parameter β) (Fig 7.2). The basic results are summarized in Fig 7.2. Consider the CTL efficacy first. If the CTL efficacy is very weak, tissue size is similar to the levels in the absence of infection (Fig 7.2). This is because the CTL hardly have any effect on the dynamics, and the virus is assumed to be noncytotoxic. That is, although infected cells are present, the total tissue size is not reduced by the infection. As the CTL efficacy is increased, the total number of tissue cells declines down to a minimum (Fig 7.2). This is because the CTL are not efficient enough to reduce virus load significantly. High virus load, combined with CTL-mediated lysis, results in pathology. As the CTL efficacy is increased further, the tissue size rises again (Fig 7.2). This is because the CTL are more efficient in this parameter region. Thus, virus load is reduced to relatively low levels, and the killing of infected cells causes less damage to the tissue. As the efficacy of the CTL response is increased further, the tissue size approaches preinfection levels (Fig 7.2) because the virus is cleared (i.e. average number of virus particles is very low).

Therefore, there is an intermediate CTL efficacy when pathology is maximized (Fig 7.2). This level of CTL efficacy is determined by the replication rate of the virus (Fig 7.2). The faster the replication rate of the virus, the higher the CTL efficacy at which pathology is maximized, and the higher the efficacy of the CTL has to be in order to overcome pathology and to clear the infection (Fig 7.2). In addition, the faster the viral replication rate, the

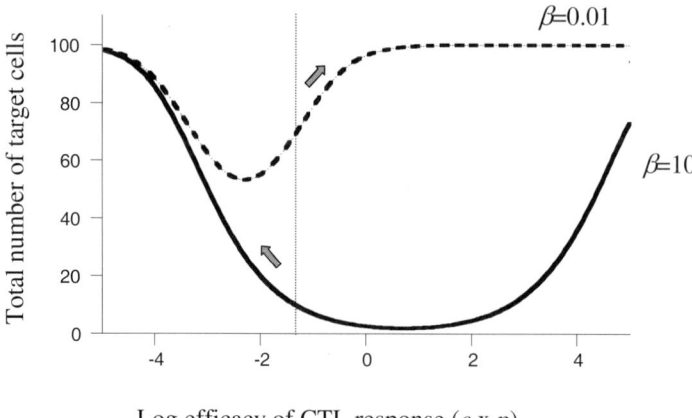

Fig. 7.2. Basic properties of CTL-induced pathology, defined by a reduction of the total number of target cells in the presence of CTL, compared to the absence of CTL. Plot is based on equation (2.11). We assume that the virus is noncytotoxic. CTL-induced pathology is most likely to occur at a low or intermediate efficacy of the CTL response (cp). In addition, the replication rate of the virus plays an important role. The faster the replication kinetics of the virus, the more severe the degree of pathology observed. If the virus replicates at a fast rate, a significant reduction in the total number of target cells will be observed even in the presence of a relatively strong CTL response. If the virus replicates slowly, any degree of immunopathology is only observed in the presence of inefficient CTL. Thus, for slowly replicating viruses, an increase in the CTL responsiveness is likely to benefit the host, while for faster replicating strains, the opposite applies (see vertical dashed line and arrows). Parameters were chosen as follows: $\lambda = 10$, $d = 0.1$, $a = 0.1$, $k = 1$, $u = 1$, $b = 0.1$.

more pronounced the degree of CTL-induced pathology (Fig 7.2). Faster viral replication correlates with more infection events, and thus with a higher number of cells that are killed by the CTL. In the context of this model, it is not possible to provide meaningful analytical expressions for the CTL efficacy at which the level of pathology is maximized, and for the degree of pathology. The results described here are based on numerical simulations.

Although we have explored the properties of CTL-induced pathology assuming a noncytotoxic virus, our results do not strictly depend on this assumption. In general, the presence of a CTL response can lead to a reduction in the total number of target cells if the cytotoxicity of the virus is low relative to the rate of viral replication. In terms of the model, CTL can contribute to tissue damage if $a < (\lambda\beta)^{1/2}$, where a denotes viral cytotoxicity and $\lambda\beta$ correlates with the replication kinetics of the virus [Krakauer and Nowak (1999)].

Therefore, if the virus replicates at a fast rate, CTL can contribute to pathology even if the virus is relatively cytotoxic.

Based on these findings we can give a definition of CTL-induced pathology. CTL-mediated pathology occurs if the sum of uninfected and infected target cells at equilibrium is smaller in the presence of a CTL response than in the absence of CTL. It is brought about by a response of intermediate strength: if the CTL are too weak they do not affect the dynamics significantly; if they are too strong, they resolve the infection successfully. If their efficacy is intermediate, the CTL fail to resolve the infection, and continuously kill infected cells at a significant rate. The degree of CTL-induced pathology is stronger the faster the replication rate of the virus. CTL-induced pathology is not only observed in noncytotoxic viruses. It can be observed as long as the viral cytotoxicity is low relative to the viral replication rate.

7.2 CTL-Induced Pathology in LCMV Infection

LCMV infection is an ideal system to test mathematical predictions. The occurrence of CTL-induced pathology, resulting from the killing of infected cells, is clearly documented with LCMV. In addition, viral and host parameters can be varied easily. According to theory, the viral replication rate and the efficacy of the CTL response are key parameters that determine whether CTL cause harm or not. Several LCMV strains exist that differ in their replication rate [Moskophidis et al. (1995b)]. In addition, the viral replication rate can be modified by knocking out IFN-γ in mice [Bartholdy et al. (2000); Nansen et al. (1999); Thomsen et al. (2000); van den Broek et al. (1995a); van den Broek et al. (1995b)]. Absence of IFN-γ increases the rate of viral replication. The rate of viral spread can be further influenced by varying the initial virus load. If mice are infected with a higher dose of the virus, the initial spread of the virus is faster. The efficacy of the CTL response can be modulated by knocking out CD4 T cell help [Andreasen et al. (2000); Borrow et al. (1996); Borrow et al. (1998); Christensen et al. (2001); Thomsen et al. (1996); Thomsen et al. (2000); Thomsen et al. (1998)]. Absence of helper cells results in a significantly reduced ability of CTL to control LCMV infection in the long-term. Therefore, experiments in IFN-γ deficient mice and CD4 cell deficient mice can be used to address the mathematical insights described above. The effect of CD4 T cell help on CTL responses is discussed in detail in Chapter 4.

First, consider the replication rate of the virus (Fig 7.3) [Nansen et al. (1999)]. We compare two strains. *LCMV Armstrong* replicates slowly, while *LCMV Traub* replicates fast. Consider wild-type mice first. With a relatively low initial virus dose of 200 PFU (plaque forming units, a measure of virus load), the virus is controlled efficiently with both LCMV Armstrong and LCMV Traub infection (Fig 7.3). If the initial virus load is a hundred fold higher, then mice infected with LCMV Traub exhibit a significant degree

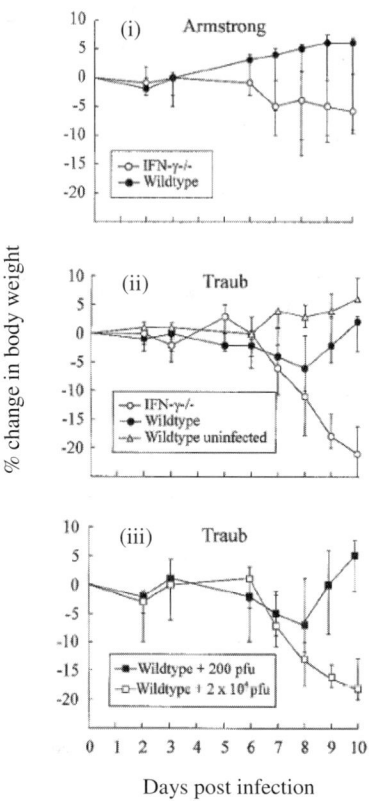

Fig. 7.3. Degree of CTL-induced pathology observed in mice infected with the slowly replicating LCMV strain Armstrong, and the fast replicating LCMV strain Traub. Wild-type mice as well as IFN-γ deficient mice were infected. Pathology is measured as percent change in body weight of the mice. Data taken from [Nansen et al. (1999)]. (i) Infection with the slowly replicating LCMV Armstrong does not cause significant pathology, even in IFN-γ deficient mice. (ii) Infection with the fast replicating Traub strain results in significant pathology in IFN-γ deficient mice. (iii) Infection with the fast LCMV Traub also results in significant pathology in wild-type mice if the infectious dose is increased.

of CTL-induced pathology (Fig 7.3). In IFN-γ deficient mice, even infection with the slow LCMV Armstrong results in limited pathology (Fig 7.3). On the other hand, infection with fast LCMV Traub strain results in severe CTL-induced pathology and death of the animals (Fig 7.3). These data confirm the prediction that faster viral replication kinetics promote the occurrence of CTL-induced pathology.

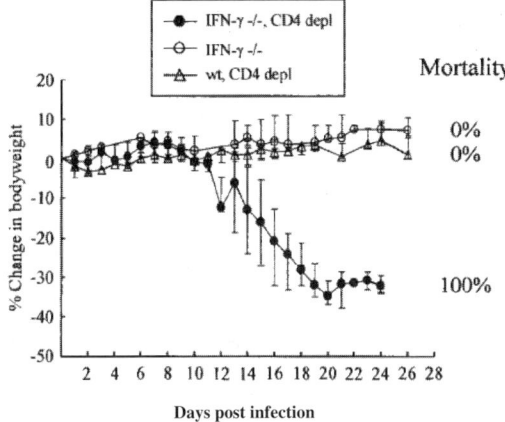

Fig. 7.4. Effect of CD4 cell deficiency (weaker CTL response) and IFN-γ deficiency on the development of CTL-induced pathology in mice infected with the slowly replicating LCMV Armstrong. Pathology is measured as percent change in body weight of mice. Data taken from [Christensen et al. (2001)]. A deficiency of CD4 T cell help or IFN-γ alone does not result in significant pathology. On the other hand, strong pathology and death of animals is observed if mice are deficient in both CD4 T cell help and IFN-γ.

Now, consider variation in the efficacy of the CTL response in addition to the variation in the viral replication rate (Fig 7.4). We consider infection of different types of mice with the slowly replicating LCMV Armstrong (Fig 7.4). Four genetically different mice were infected [Christensen et al. (2001)]: wild-type mice (CD4+ and IFN-γ+) that have an efficient CTL response and can reduce the rate of viral replication; CD4+ IFN-γ mice that have an efficient CTL response, but viral replication is faster due to the absence of IFN-γ; CD4- IFN-γ+ mice that have a reduced efficacy of the CTL response, but can still reduce the rate of viral replication through IFN-γ; and CD4- IFN-γ- mice that have a weak CTL response and allow the virus to replicate faster. Infection of wild-type mice induces a potent CTL response that rapidly controls the infection. Pathology is not observed (Fig 7.4). A deficiency in either CD4 cell help or IFN-γ alone also does not result in significant degrees of pathology in the long-term (Fig 7.4). On the other hand, a deficiency in both CD4 cell help and IFN-γ results in severe CTL-induced pathology and death of the mice (Fig 7.4). The pathology was shown to be dependent on CTL-mediated antiviral activity. This set of experiments confirms the prediction that the efficacy of the CTL response (modulated through CD4 cell help) is also an important determinant of pathology, and that pathology is most likely to be observed if the replication rate of the virus is fast relative to the efficacy of the CTL response. As suggested by the model, CTL-mediated pathology is the consequence of a relatively weak, and not a strong response.

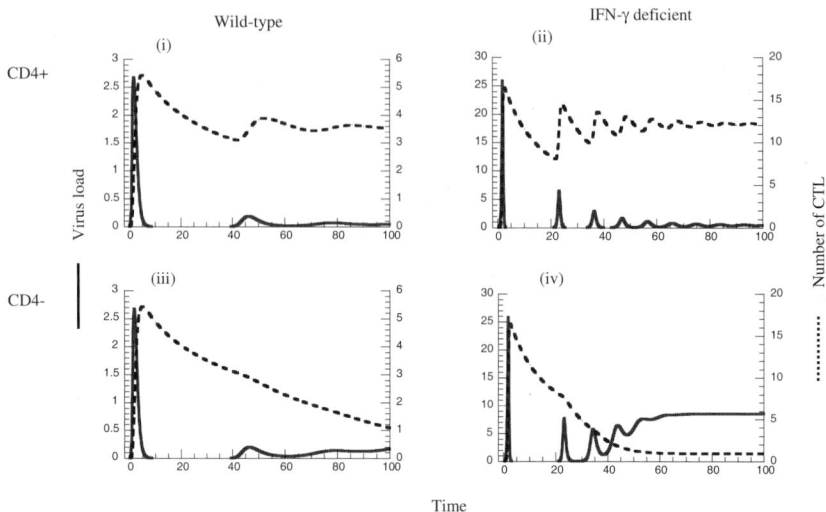

Fig. 7.5. Effect of CD4 cell help and IFN-γ deficiency on the dynamics of LCMV and the LCMV specific CTL response, as predicted by the mathematical model (7.1–7.4). (i) wild-type hosts; (ii) IFN-γ-/- but helper competent hosts; (iii) IFN-γ competent but helper deficient hosts; (iv) hosts deficient in both IFN-γ and CD4 cell help. For details see text. Parameters were chosen as follows: $\lambda = 10$, $d = 0.1$, $a = 0.1$, $p = 1$, $c = 1$, $b = 0.01$, $g = 0.1$, $h = 0.1$, $\beta = 0.05$, $s = 1$, $q = 100$. IFN-γ-/- hosts are characterized by $q = 0$, while helper deficient hosts are characterized by $s = 0.01$. Note the different scales on the y-axis for wild-type and IFN-γ deficient hosts.

These experiments with the slowly replicating LCMV Armstrong strain can in fact be simulated using a variation of the basic CTL dynamics model discussed earlier in this chapter (2.2–2.4, 2.11) [Christensen et al. (2001)]. It includes IFN-γ mediated suppression of viral replication and CD4 T cell help (Fig 7.5). The model is briefly outlined as follows. It consists of four variables: the population of uninfected cells x, infected cells y, CTL precursors w, and CTL effectors z. It is given by the following set of equations:

$$\dot{x} = \lambda - dx - \frac{\beta xy}{qz+1}, \tag{7.1}$$

$$\dot{y} = \frac{\beta xy}{qz+1} - ay - pyz, \tag{7.2}$$

$$\dot{w} = csyw - cgyw - bw, \tag{7.3}$$

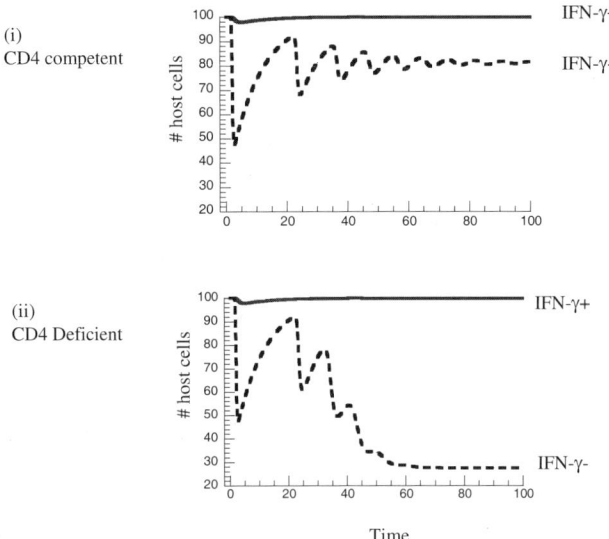

Fig. 7.6. Effect of CD4 helper cell deficiency and IFN-γ deficiency on the level of CTL-induced pathology, as predicted by the mathematical model (7.1–7.4). (i) IFNγ-/- hosts with an intact CD4 helper cell response. (ii) IFNγ-/- hosts deficient in CD4 cell help. For details see text. Parameters were chosen as follows: $\lambda = 10$, $d = 0.1$, $a = 0.1$, $p = 1$, $c = 1$, $b = 0.01$, $g = 0.1$, $h = 0.1$, $\beta = 0.05$, $s = 1$, $q = 100$; IFNγ-/- hosts are characterized by $q = 0$, while helper deficient hosts are characterized by $s = 0.01$.

$$\dot{z} = cgyw - hz. \tag{7.4}$$

The equations are based on the mathematical models for CTL responses described in chapter 2 (2.12–2.13), and include two additions. First, the basic rate of virus replication can be reduced by IFN-γ, secreted by CTL, with a rate q. Second, the rate of CTL proliferation is dependent on help, which is captured in the parameter s. For more general models of nonlytic inhibitors secreted by CTL and models of CD4 T cell impairment, see Chapters 8 and 11, respectively.

The simulation results are as follows. In the wild-type scenario (Fig 7.5i), a sustained CTL memory response is generated that resolves the primary infection and ensures long-term immunological control and clearance. The IFN-γ deficient scenario (Fig 7.5ii) is characterized by low level persistent virus replication. However, the CTL response is still sustained, and controls the infection over the long-term. The higher virus load maintains a higher number of CTL in the memory phase compared with the wild-type scenario, and small bursts of virus drive the CTL number up, which can keep viral replication in check. The validity of this prediction has been documented by experiments [Bartholdy et al. (2000)]. The CD4 helper deficient scenario (Fig 7.5iii) also results only in a small loss of virus control. However, the simulation suggests that over time the CTL response slowly decays and this can lead to reduced levels of virus control over the long-term. Consistent with this, a slow decline in the number of CTL has been observed in CD4 cell deficient mice infected with LCMV Armstrong [Christensen et al. (2001)]. In the scenario that is characterized by both a CD4 cell deficiency and an IFN-γ deficiency (Fig 7.5iv), there is a rapid loss of virus control and a significant reduction in the level of the CTL response. These patterns are also reflected in Fig 7.6, which shows model predictions about the effect of CD4 cell deficiency and IFN-γ deficiency on the level of CTL-induced pathology (characterized by a low number of host tissue cells). In the presence of CD4 cell help, the model suggests that the host experiences transient pathology, followed by relatively efficient virus control. In accordance with the experiments, IFN-γ deficiency does not result in significant levels of pathology in the long-term. In the absence of CD4 cell help, the simulations suggests that while no significant long-term pathology is observed in the presence of IFN-γ, the absence of IFN-γ leads to severe CTL-induced pathology, which corresponds to the wasting and death of hosts (Fig 7.6). Thus, the predictions from computer simulations coincide with the experimental results described above.

7.3 CTL-Induced Pathology and HIV Infection

Based on his work on immunopathology in LCMV infection, Rolf Zinkernagel suggested that T cell depletion in HIV infection might not be brought about by the virus itself, but by the CTL response that kills infected T helper cells [Zinkernagel (1994); Zinkernagel (1995); Zinkernagel and Hengartner (1994); Zinkernagel et al. (1999)]. He argued that HIV might be noncytotoxic and that AIDS is the consequence of CTL-induced pathology. Consequently, treatment could be aimed at reducing the CTL responses against HIV. This is a controversial concept, and this section explores this topic from a mathematical perspective. Clinical data indicate that HIV is cytotoxic, especially during the later stages of the disease. Nevertheless, the basic modeling described above has shown that CTL can contribute to tissue pathology even in the context of cytotoxic viruses if the viral replication rate is sufficiently fast. However, HIV brings additional complexity. We have to explain the depletion of the entire T helper cell population, although HIV only infects a fraction of all available T helper cells. In particular, HIV tends to infect activated T cells preferentially [Heinzinger et al. (1994); Stevenson (1994); Stevenson (1996); Stevenson et al. (1995)]. The basic model of infection dynamics (2.2–2.4) can be modified to account for this, and we will discuss the conditions under which CTL-induced pathology can occur in this HIV specific scenario. The model is described as follows.

7.3 CTL-Induced Pathology and HIV Infection

The model consists of five variables: resting uninfected CD4 T cells s, activated uninfected T cells x, infected T cells y, free virus v, and CTL z. In contrast to the basic virus infection model we assume that HIV only completes it replication cycle in activated CD4 T cells and that T cell activation occurs in response to viral antigen [Wodarz and Krakauer (2000); Wodarz et al. (1999)]. This can include bystander activation resulting from high viral loads. The model is given by the following set of differential equations.

$$\dot{s} = \xi - fs - \frac{rsv}{x + \epsilon}, \quad (7.5)$$

$$\dot{x} = \frac{rsv}{x + \epsilon} - dx - \beta xy, \quad (7.6)$$

$$\dot{y} = \beta xy - ay - pyz, \quad (7.7)$$

$$\dot{v} = ky - uv, \quad (7.8)$$

$$\dot{z} = cy - bz. \quad (7.9)$$

Resting CD4 T cells are produced at a rate ξ, die at a rate fs and become activated by virus at a rate $rsv/(x + \epsilon)$. This assumes that the rate of T cell activation is a function of the number of T cells that are already activated. If ϵ is small, the rate of T cell activation significantly increases if the number of activated T cells is low. On the other hand, if the value of ϵ is large, then the rate of T helper cell activation does not depend on the number of already activated cells, but only on virus load. Activated T cells die at a rate dx and become infected at a rate βxv. Infected cells produce free virus at a rate ky, die at a rate ay and are killed by CTL at a rate pyz. Free virus decays at a rate uv. The CTL population expands in response to antigen at a rate cy and decays at a rate bz. Persistent virus replication in the presence of a CTL response is described by an equilibrium given by a third degree polynomial expression, so results are obtained by numerical simulations.

So far, we measured pathology by the number of infected plus uninfected cells. This is still true in the current context, but we have to consider both the resting and the activated cells. Therefore, we determine by how much the populations of resting uninfected T cells, activated uninfected T cells, and infected T cells $(s + x + y)$ have been reduced compared to preinfection levels.

The exact mathematical consequences of assuming that only activated T cells become infected are set out in detail in [Wodarz et al. (1999)]. Here we are interested in whether CTL-mediated pathology can lead to a depletion of the overall CD4 T cell count at equilibrium. In the previous section we found that CTL-induced pathology is the consequence of fast viral replication relative to the CTL responsiveness of the host. In general, this also holds true in the HIV specific model: fast virus replication relative to the strength of the CTL response can deplete the population of CD4 T cells, even if the virus only infects the subpopulation of activated T cells. This is true especially for a fast rate of virus production by infected cells k. A faster rate of virus production results in a higher number of free virions exposed to the immune system.

This in turn results in the activation of more CD4 T cells from the pool of resting cells. This makes more cells available for infection and for killing. If more cells become activated and infected, and if the infected cells become killed by the CTL, we observe CTL-mediated depletion of a large fraction of the CD4 T cell population. Virus replication, however, consists of more components than the rate of virus production, most importantly the rate of virus entry into susceptible cells β. While in the general virus dynamics model (2.2–2.4), this distinction makes no difference, in the HIV specific model a fast rate of viral entry into susceptible cells may or may not contribute to the the development of CTL-induced pathology. This is because a fast rate of viral entry into susceptible cells β does not lead to a significantly higher activation rate of CD4 T cells and therefore does not promote the killing of more infected CD4 T cells. A fast rate of virus entry can only contribute to CTL-induced T cell depletion if the rate of T cell activation increases significantly when the number of functional (uninfected) T cells falls to low levels through a feedback mechanism (low value of ϵ). In this case, a high value of β depletes the population of activated T cells, and the feedback mechanisms indirectly induces a higher rate of CD4 T cell activation. Whether such a feedback mechanism exists or not is unclear.

To summarize, a fast rate of HIV replication can contribute to the development of AIDS through CTL-induced pathology, even if the virus can only infect the subpopulation of activated CD4 T cells. This is because a fast rate of viral replication can lead to an accelerated activation rate of the CD4 T cells. Consequently more cells are available for infection and killing. This mechanism also holds if HIV itself contributes to cell death, as documented by data.

If CTL-induced pathology can contribute to the development of AIDS, can a depletion of the CTL response have a therapeutic benefit? For the sake of the argument, let us assume that HIV is noncytotoxic. In this case, the basic modeling discussed above, and experiments with LCMV, suggest that pathology can be avoided either if the CTL response is entirely absent, or if it is strong relative to the replication rate of the virus (if the virus was cytotoxic, then the absence of CTL would not prevent tissue pathology). In the case of the HIV model however, the situation is different. The answer depends on the life span of activated relative to resting CD4 T cells. For fast viral replication kinetics, the model predicts that the sum of infected and uninfected T cells in the absence of a lytic CTL response is approximately ξ/d; that is the production rate of resting cells divided by the death rate of activated T cells. In the absence of the infection, the total number of CD4 T cells is given by ξ/f, where f is the death rate of resting CD4 T cells. Thus, under the reasonable assumption that the death rate of activated T cells is significantly higher than that of resting T cells ($d >> f$), a fast rate of viral replication in the absence of a CTL response can still produce a reduction of the CD4 T cell count. For example, the CD4 T cell count can be reduced from

1000 to 200 if the death rate of activated cells is five times higher than that of resting T cells. Therefore, depletion of the CTL response is not likely to have a therapeutic benefit. Instead, therapy should boost CTL responses so they are sufficiently strong relative to the replication rate of the virus. Then, CTL-induced pathology does not occur.

7.4 Summary

This chapter examined mathematically the conditions under which antiviral CTL responses are expected to give rise to pathology instead of resolving the infection. CTL-induced pathology is promoted by a high rate of viral replication relative to the efficacy of the CTL response. Contrary to some suggestions, CTL-induced pathology is not promoted by a strong response, but by a response that fails to reduce the virus to low levels. High virus load, coupled with ongoing lysis of infected cells, results in tissue depletion. This framework was applied to the analysis of LCMV infection, and predictions are strongly supported by experimental studies. The mathematical models further suggest that CTL-induced pathology can contribute to the depletion of T helper cells in HIV infection, even if the virus is cytotoxic. In contrast to some arguments, however, the model also suggests that the elimination of the CTL response would not have any beneficial effect for the patient, and that the CTL response should instead be boosted by treatment.

8
Lytic versus Nonlytic Activity

Chapter 7 discussed the concept of CTL-induced pathology, where CTL-mediated lysis of infected cells can lead to tissue damage and mortality of the host, instead of virus control. A fast rate of viral replication relative to the efficacy of the CTL was found to promote the occurrence of CTL-induced pathology. Lysis of infected cells, however, is not the only mechanism by which CTL can fight viral infections (Fig 8.1). CTL can also secrete soluble factors that bind to infected cells [Guidotti et al. (1994a)]. This triggers a reaction inside the cell that inhibits virus replication (Fig 8.1). Consequently, while the cell remains infected and is not killed, it stops producing virus particles and does not contribute to virus spread anymore. This type of mechanism is called *nonlytic activity* of CTL. There are examples of this mode of CTL-mediated activity across several infections. In LCMV infection, IFN-γ, secreted by CTL, can reduce the rate of viral replication [Bartholdy et al. (2000)]. In HIV infection, a soluble factor has been identified that can stop virus production by infected cells. It has been termed CTL secreted antiviral factor (CAF), and its exact identity if subject to debate [Levy et al. (1996); Zhang et al. (2002)]. In addition, HIV specific CTL can secrete chemokines that can inhibit the entry of certain HIV strains into their target cells [Cocchi et al. (1995); Gallo (1997); Moore et al. (1997); Scarlatti et al. (1997)]. nonlytic CTL responses are thought to play an important role in HBV infection [Guidotti et al. (1994b); Guidotti et al. (1999a); Guidotti and Chisari (1996); Guidotti et al. (1996a); Guidotti et al. (1999b)]. The ability of CTL to silence the virus inside infected cells may be crucial for the successful control of the virus without the occurrence of liver pathology. It has also been reported that in HBV infection, soluble factors secreted by CTL, might destruct the viral genome in the infected cells, returning them to an uninfected state. The CTL therefore cure the cells rather than killing them. This chapter examines the relative role of lytic versus nonlytic CTL responses. Which type of response is required for the resolution of different infections? Is there a certain balance of lytic and nonlytic immunity, which is needed to fight viral infections?

8 Lytic versus Nonlytic Activity

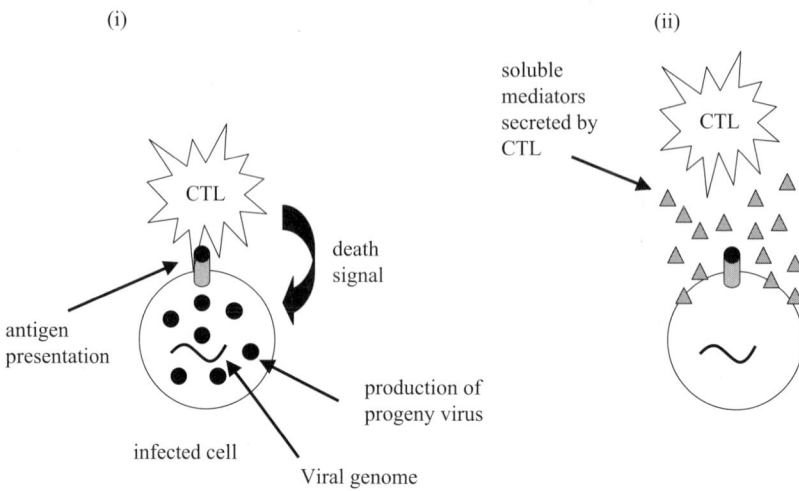

Fig. 8.1. Two different modes of CTL-mediated activity. (i) CTL-mediated lysis involves specific contact between the CTL and the infected cell, and this results in a death signal and apoptosis of the infected cell. (ii) nonlytic CTL activity involves the secretion of soluble mediators by CTL. These soluble mediators influence the infected cell such that virus replication in the cell is shut down. The infected cell remains alive, but does not produce virus anymore. There are variations of this concept. For example, the soluble mediators can prevent infection of the cell, or even lead to the removal of the viral genome from the cell. With all forms of nonlytic activity, direct contact between the CTL and the infected cell is not required, because the soluble mediators are released and can find the infected cell.

Many experiments have addressed the important question of which types of immune mediators are responsible primarily for controlling specific virus infections [Bartholdy et al. (2000); Kagi and Hengartner (1996); Kagi et al. (1995a); Kagi et al. (1996); Kagi et al. (1995b)]. In this respect, genetic knockout mice that lack the ability to perform CTL-mediated lysis, or that lack the ability to perform nonlytic activity, have been useful. Lysis can be achieved in two basic ways. The most important mode is the use of *perforin* molecules. These are secreted by CTL and induce apoptosis of the infected cell. A less important mode of lysis occurs through the interaction between *FAS* and *FAS ligand*. This again triggers apoptosis of the infected cell. Both perforin- and FasL knockout mice have been used to study how CTL responses deal with viral infections in the absence of lysis. On the other hand, IFN-γ knockout mice can be used to study how reduced nonlytic activity can influence the ability of CTL to control viral infections. Note that nonlytic CTL responses act in a similar way compared to antibody responses. They do not result in the death of infected cells, but inhibit the spread of the virus from cell to cell. Therefore, in order to determine how the relative balance of lytic and nonlytic responses affects the ability of CTL to control viral infections, antibody deficient mice have been used to reduce the overall nonlytic activity of the immune system. Such data will also be discussed in this chapter.

The following results and suggestions have been put forward by experimental work. Perforin knockout mice do not lose control of certain cytopathic viruses [Kagi et al. (1996); Kagi et al. (1995b)]. However, perforin knockout mice infected with the noncytopathic lymphocytic choriomeningitis virus (LCMV) have been reported to be severely compromised in their ability to control the infection [Kagi et al. (1996); Kagi et al. (1995b)]. Based on these observations, Kagi and others formulated the hypothesis that CTL-mediated lysis is an essential immune mechanism for fighting noncytopathic viruses in general, whereas soluble immune factors are sufficient to combat cytopathic viruses [Kagi et al. (1996); Kagi et al. (1995b)]. The explanation was that cytopathic viruses kill the infected cells themselves. Thus, lytic activity is not required to remove these viral reservoirs. On the other hand, lytic activity is required for noncytopathic viruses because they do not kill the infected cells. However, other and subsequent experiments have demonstrated that the situation is more complicated than this. Soluble factors have been shown to contribute to the resolution of noncytopathic infections [Bartholdy et al. (2000); Ciurea et al. (2000); Guidotti et al. (1994b); Guidotti and Chisari (1996); Guidotti et al. (1996b); Nansen et al. (1999); Planz et al. (1997); van den Broek et al. (1995a); van den Broek et al. (1995b)], and some cytopathic infections can be cleared independently by soluble and lytic effectors [Eichelberger et al. (1991); Topham et al. (1997)].

The effect of lytic and nonlytic immune responses on the dynamics between replicating viruses and the immune system has been studied by mathematical models. This chapter summarizes insights gained from these models and compares them to experimental data. The mathematical models suggest that a correct balance between lytic and nonlytic immunity is crucial to achieve virus control, and that a shift in this balance can lead to loss of control and immunopathology.

8.1 Modeling Lytic and Nonlytic CTL Responses

Here we describe a modification of the basic model of virus dynamics (2.2–2.4) and of the CTL response equation (2.11) to include both lytic and nonlytic CTL activity [Wodarz et al. (2002)]. It includes the usual variables: susceptible host cells x, a virus population y, and immune responses z. Susceptible host cells are generated at a rate λ, die at a rate dx and become infected by virus at a rate βxy. Virus replication is inhibited by the immune response at a rate $qz + 1$. This corresponds to nonlytic antiviral activity. It is assumed that nonlytic immunity can in principle affect any stage of viral replication, e.g. virus production or infectivity. Infected cells die at a rate ay and become killed by the immune system at a rate pyz. This corresponds to lytic effector mechanisms. The immune response is assumed to get stronger at a rate proportional to the number of infected cells cy, and also decays exponentially at a rate proportional to its current strength bz. Note that the variable z represents overall immunity that can be generated in response to a virus infection. The parameter p expresses the strength of the lytic component, whereas the parameter q expresses the efficacy of the nonlytic component. The model is described by the following set of differential equations.

$$\dot{x} = \lambda - dx - \frac{\beta xy}{qz+1}, \qquad (8.1)$$

$$\dot{y} = \frac{\beta xy}{qz+1} - ay - pyz, \qquad (8.2)$$

$$\dot{z} = cy - bz. \qquad (8.3)$$

Te basic conditions when an infection and the CTL response can become established are identical to those discussed in Chapter 2 for the models (2.2–2.4) and (2.11). In the presence of a CTL response, the dynamics converge to the following equilibrium.

$$x^* = \frac{ba + cq\lambda + pbz^*}{cqd + b\beta},$$
$$y^* = bz^*/c,$$
$$z^* = \frac{-(dqca + b\beta a + dcp) + \sqrt{(dqca + b\beta a + dcp)^2 - 4pc(dcq + b\beta)(da - \lambda\beta)}}{2p(dqc + b\beta)}.$$

In terms of the clinical outcome of the infection, two measures have to be taken into consideration. The first measure is the equilibrium virus load y^*. The higher the virus load, the less efficient the immune system is in limiting viral replication. Very low virus loads correspond to immune-mediated control or clearance. The second measure is the total number of host cells, uninfected plus infected, at equilibrium $x^* + y^*$. Even if the number of infected cells is reduced to relatively low levels, depletion of the population of uninfected host cells can result in tissue damage. Hence, resolution of the infection requires two conditions: (i) virus load has to converge toward low levels. (i.e. $y^* \to 0$); (ii) the total number of host cells has to converge to the preinfection level. (i.e. $x^* + y^* \to \lambda/d$).

8.2 Effect of Lytic and Nonlytic Immunity on Virus Control

Lytic and nonlytic effector mechanisms influence two basic viral properties: the death rate of infected cells a and the replication rate of the virus β. To

8.2 Effect of Lytic and Nonlytic Immunity on Virus Control

understand the relevance of immune responses, it is useful to consider the effect of these two basic parameters on virus load and the total number of target cells in the absence of immune control. The higher the death rate of infected cells a, the lower the virus load (Fig 8.2). If the value of a lies above a threshold, then the basic reproductive ratio of the virus R_0 is less than unity, and the infection becomes extinct. By contrast, an increase in the viral replication rate β results in an asymptotic increase in virus load (Fig 8.2). If the value of β lies below a threshold, then $R_0 < 1$ and the infection becomes extinct. Hence, both a decrease in the viral replication kinetics due to nonlytic effector mechanisms and an increase in the death rate of infected cells due to lytic effectors can contribute to a reduction in R_0 and virus load.

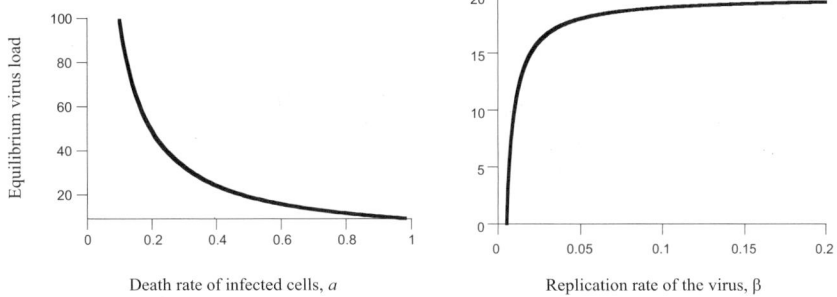

Fig. 8.2. Effect of the death rate of infected cells a and the replication rate of the virus β on equilibrium virus load, y^*, according to equations (8.1–8.3). The faster the death rate of infected cells, and the slower the rate of viral replication, the lower the virus load. Baseline parameters were chosen as follows: $\lambda = 10$, $d = 0.1$, $a = 0.5$, $\beta = 0.1$.

The effect of these viral parameters on the total number of host cells is more complex (Fig 8.3). The total number of host cells is not a monotonic function of the death rate of infected cells a. Increasing the parameter a first leads to a decrease of the total number of host cells down to a minimum. A further increase in a results in an increase in the total number of host cells until a crosses a threshold where $R_0 < 1$. Then, the virus is extinct. The minimum equilibrium number of host cells is given at a death rate $a = (\lambda \beta)^{1/2}$ and it has a value of $[x * + y*]_{\min} = 2(\lambda/\beta)^{1/2} - d/\beta$. Thus, the higher the rate of viral replication, the higher the death rate of infected cells at which the total number of host cells reaches the minimum, and the lower the value of this minimum. (Fig 8.3). For very slow rates of viral replication, the minimum number of host cells approaches a value of λ/d, which is equivalent to the number of host cells in the absence of an infection. These observations give rise to two basic results. Reducing the replication rate of the virus by nonlytic effector mechanisms is always beneficial to the host. Increasing the

118   8 Lytic versus Nonlytic Activity

death rate of the infected cells by lytic effector mechanisms can be both detrimental and beneficial to the host. Lytic effectors are likely to be detrimental if the virus replicates at a fast rate. In the following sections, we examine the relative contribution of lytic and nonlytic effector mechanisms in noncytopathic and more cytopathic infections. We apply these findings to specific cases to interpret experimental data.

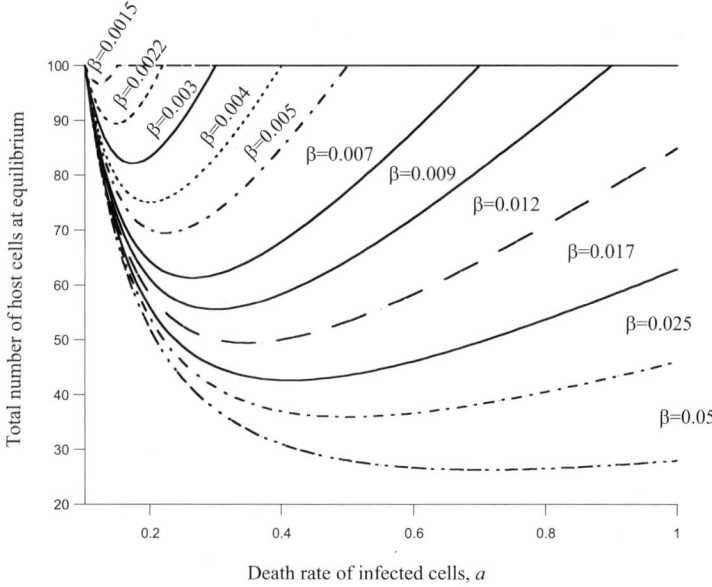

Fig. 8.3. Effect of the death rate of infected cells a and the replication rate of the virus β on the total number of host cells at equilibrium $x^* + y^*$ (a measure of CTL-induced pathology). The total number of host cells is lowest (and pathology is highest) for intermediate death rates of infected cells. The faster the replication kinetics of the virus, the lower the minimum number of host cells, and the higher the death rate of infected cells at which this minimum is attained. The plot is based on equations (8.1–8.3). Baseline parameters were chosen as follows: $\lambda = 10$, $d = 0.1$, $a = 0.5$. The viral replication rates for the individual curves are shown in the figure.

8.3 Noncytopathic Viruses

In terms of our model, we define a noncytopathic virus by assuming $a = d$. That is, the death rate of infected cells in the absence of immunity equals

8.3 Noncytopathic Viruses 119

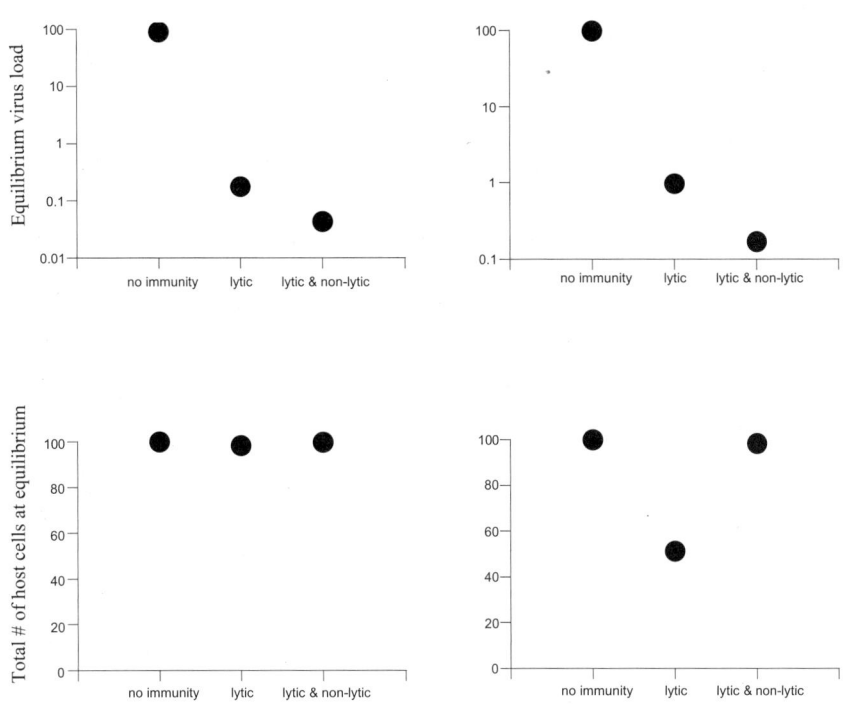

Fig. 8.4. Control of noncytopathic viruses by lytic and nonlytic effector mechanisms. (i) Slowly replicating virus. Lytic effectors alone can achieve a similar level of virus control as a combination of lytic and nonlytic effectors. (ii) Fast replicating virus. Lytic effectors alone cannot control the infection, because it results in immunopathology. Cooperation of lytic and nonlytic effector mechanisms can control the infection, because nonlytic immunity slows down the overall replication kinetics of the virus. Note that the effect of a nonlytic response alone has not been plotted. This is because we consider noncytopathic viruses that do not kill their target cells. Since the life span of infected cells is not reduced, absence of lysis is unlikely to result in virus control in a short period of time. Plots are based on equations (8.1–8.3). Parameters were chosen as follows: $\lambda = 10$, $d = 0.1$, $a = 0.1$, $q = 10$, $p = 1$, $c = 0.1$, $b = 0.1$. For (a) $\beta = 0.01$; For (b) $\beta = 0.1$.

that of uninfected target cells. In this case, the absence of lytic immune effectors could result in failure to clear the infection, but because the virus is noncytopathic, this might not harm the host in clinical terms. An example is CTL exhaustion in LCMV infection [Moskophidis et al. (1993c)]. On the other hand, prolonged persistence of replicating antigen at high loads could lead to a chronic state of activation of the immune system, resulting in the overproduction of cytokines and hence damage to the host. This is observed when perforin-deficient mice are infected with LCMV [Binder et al. (1998)]. These mice develop a severe cytokine-mediated aplastic anemia, and may succumb from the infection. Another possible example of this outcome is human T cell leukemia virus (HTLV-1) infection [Bangham (1993)]. In addition, noncytopathic viruses can induce cell damage without killing, for example by turning off so called 'luxury' functions [Oldstone et al. (1982)]. Therefore, evolutionary pressure will have favored lytic effector mechanisms to clear noncytopathic viruses. However, according to the model, it will be hard for lytic effectors to resolve an infection with a noncytopathic virus, especially if it is replicating at a fast rate (Fig 8.3). The lytic effectors increase the death rate of infected cells and thus can result in the depletion of target cells and immunopathology. The faster the replication rate of the virus, the stronger the CTL response has to be to minimize immunopathology and result in resolution of the infection (Fig 8.3). According to the model, this difficulty can be overcome by combined action of lytic and nonlytic effectors. Soluble mediators reduce the replication kinetics of the virus. If the viral replication kinetics are diminished, the lytic effectors are likely to result in resolution of the infection instead of immunopathology.

Fig 8.4 summarizes the roles of lytic and soluble mediators in the clearance of noncytopathic viruses. If the virus replicates at a slow rate, lytic effectors alone are likely to be sufficient to control the infection. Although the model indicates that the absence of soluble factors could result in a slight increase in virus load, the infection is likely to be controlled in the long-term. If the virus replicates at a fast rate, soluble immune mediators are required to reduce the replication kinetics of the virus, enabling the lytic effector mechanisms to have a beneficial effect on the host.

The most extensively characterized noncytopathic virus is murine LCMV. The soluble cytokine IFN-γ has been shown to reduce the rate of viral replication in LCMV infection [Klavinskis et al. (1989); van den Broek et al. (1995a)]. Yet, the role of IFN-γ in the control of LCMV has been only worked out recently. Early studies suggested [Kagi and Hengartner (1996); Kagi et al. (1995b)] that in LCMV infection, IFN-γ did not contribute substantially to virus control. However, this conclusion was based on the analysis of mice infected with the slowly replicating Armstrong strain. More recent analysis, by contrast, has provided clear evidence for an important role for IFN-γ and other soluble mediators in the resolution of LCMV infection [Bartholdy et al. (2000); Ciurea et al. (2000); Nansen et al. (1999); Planz et al. (1997); van den Broek et al. (1995a); van den Broek et al. (1995b)]. Even with the slowly replicating Armstrong strain, analysis revealed that although IFN-γ-deficient mice infected with this strain did not show symptoms of disease, the infection was not controlled completely and significant levels of virus could be demonstrated in spleen and lungs months after infection [Bartholdy et al. (2000); Thomsen et al. (2000)]. In these mice, an equilibrium was established describing persistent LCMV replication controlled by an ongoing CTL response. Because virus load was kept at relatively low levels, pathology was virtually absent [Bartholdy et al. (2000); Nansen et al. (1999)]. This observation is in agreement with the theory presented here. Because the virus replicates at a slow rate, the model predicts that lack of nonlytic effector mechanisms will only result in a small loss of virus control and lack of severe immunopathology. The situation is different with faster replicating LCMV strains. IFN-γ-deficient mice infected with LCMV Traub quickly lose control of the infection despite the presence of efficient lytic effector mechanisms. In contrast to wild-type mice, a relatively large fraction of infected IFN-γ -/- mice succumbed to immunopathology caused by a lytic CTL response [Nansen et al. (1999)]. This is, again, in agreement with theoretical predictions (Fig 8.3). Because the virus replicates at a fast rate, soluble factors are required to significantly slow down the replication kinetics of the virus to avoid immunopathology. The absence of soluble mediators augments CTL-induced tissue damage and death of the host.

8.4 More Cytopathic Viruses

The last section examined the case of a noncytopathic virus, for which the life span of an infected cell is equal to that of an uninfected cell. Similar considerations apply to more cytopathic viruses as long as the life span of infected cells lies below a threshold, given by $a < (\lambda\beta)^{1/2}$. Within this parameter region, an increase in the death rate of infected cells can have a detrimental effect on the host (Fig 8.3). However, this does not apply if the death rate of infected cells lies above the threshold given by $a > (\lambda\beta)^{1/2}$. In this case, both lytic and nonlytic effector mechanisms always have a beneficial effect on the host. The reason is that in this parameter region, an increase in the death rate of infected cells always results in an increase in the total number of host cells (Fig 8.3). However, an effect of lytic effectors will be only apparent if the rate of immune-mediated cell killing is fast relative to the rate of virus-induced cell killing (i.e. if $pz >> a$). In this case, there are three possible scenarios regarding the role of lytic and nonlytic effector mechanisms for controlling the infection (Fig 8.5). (i) If the lytic effector mechanisms are sufficiently strong, they can resolve the infection on their own. (ii) If the nonlytic effector mechanisms are strong enough, they can also resolve the infection on their own. (iii) If neither lytic nor nonlytic effectors are sufficiently strong to resolve the disease, a combination of both mechanisms is required to overcome the infection.

An example of a virus infection characterized by a very high degree of cytopathicity is VSV infection in mice [Andersen et al. (1999); Andreasen et al. (2000); Bachmann et al. (1997); Kagi and Hengartner (1996); Thomsen et al. (1997)]. The role of different types of immune effector mechanisms in this extreme case can be investigated in MHC class I and II deficient hosts. Class

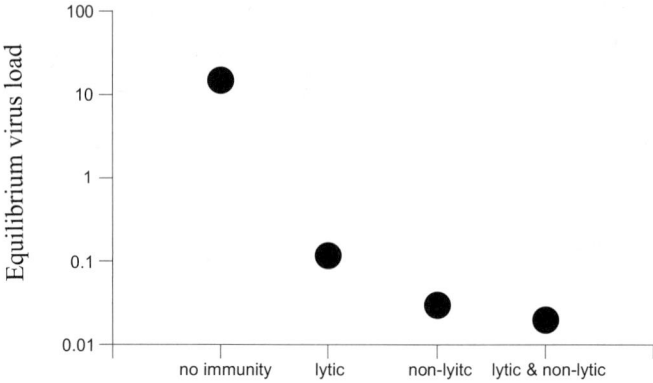

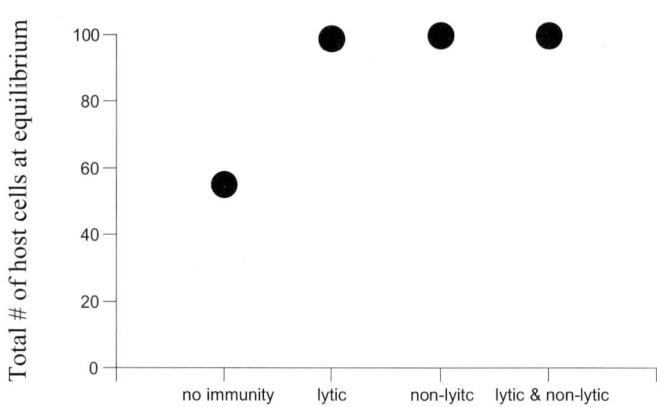

Fig. 8.5. Control of a virus characterized by a high degree of cytopathicity relative to its replication rate. Both lytic and nonlytic effector mechanisms can in principle independently control the infection. Plots are based on equations (8.1–8.3). Parameter values were chosen as follows: $\lambda = 10$, $d = 0.1$, $a = 0.4$, $\beta = 0.01$, $q = 10$, $p = 1$, $c = 0.1$, $b = 0.1$.

II-deficient mice have reduced levels of survival. However, a fraction of the infected animals survived. In these cases, CD8+ T cell responses seem to be able to control the infection partially [Andersen et al. (1999); Thomsen et al. (1997)]. In class I-deficient mice, CD4+ T cell-dependent responses can resolve the infection successfully, most probably via the induction of IgG antibody [Andersen et al. (1999); Thomsen et al. (1997)]. Animals deficient in both class I and class II cannot control VSV infection at all, resulting in the death of all mice infected [Andersen et al. (1999); Thomsen et al. (1997)]. These results are consistent with the theoretical framework presented here. VSV is extremely cytopathic, and hence the rate of virus-induced cell death could be greater than the rate of CTL-mediated lysis. Therefore, although theory suggests that, in principle, both lytic and nonlytic effector mechanisms should be able to control the infection independently, lytic mechanisms will have limited capabilities if the rate of viral cytopathicity is very high. Consequently, in class II-deficient mice, CTLs are expected to have a limited effect on virus control at best. However, we have to be cautious in the exact interpretation of the data, because CD8+ T cells could also act in a nonlytic fashion.

An example of a cytopathic virus that is not at the extreme end of the cytopathicity spectrum is influenza infection in mice [Doherty et al. (1997)]. Recovery from murine influenza virus infection has been shown to require intact T cell responses [Doherty et al. (1997); Gerhard et al. (1997); Moskophidis and Kioussis (1998)]. More specifically, experiments have revealed that both CD4+ and CD8+ T cells can promote recovery through independent mechanisms [Doherty et al. (1997); Gerhard et al. (1997)]. In the absence of CD8+ T cells, the infection can be resolved by a CD4+ T cell-dependent antibody response [Eichelberger et al. (1991)]. Absence of CD4+T cells or B cells also does not result in loss of virus control [Topham et al. (1996); Tripp et al. (1995)]. Experiments have shown that CD8+ T cells can resolve the infection through a lytic mechanism, mediated either by perforin or FAS [Topham et al. (1997)]. The result that both lytic and nonlytic effector mechanisms can independently clear influenza infection in mice is in agreement with the theoretical considerations presented here. Because the virus is cytopathic, both a sufficient increase in the death rate of infected cells and a decrease in the rate of viral replication are expected to have a beneficial effect on the host and lead to resolution of the disease. With cytopathic viruses, a collaboration between both types of effector is less likely to be required to ensure resolution of the disease, especially if the virus challenge is not overwhelming.

8.5 Summary

We have used mathematical models to analyze the role of lytic and nonlytic effector mechanisms in viral infections. Theory, complementing experimental data, argues against the simple rule that lytic effectors are required to deal with noncytopathic viruses, whereas soluble factors are sufficient to deal with cytopathic viruses. Instead, the following pattern is suggested. In the present context, the distinction between noncytopathic and cytopathic viruses is not precise. The relevance of lytic and nonlytic effector mechanisms for resolving the infection depends on the viral cytopathicity relative to the rate of viral replication. If cytopathicity lies below the threshold defined by $a < (\lambda\beta)^{1/2}$, then a combination of lytic and nonlytic effector mechanisms is likely to be required to resolve the disease. The higher the replication rate of the virus, the more nonlytic effectors are required to resolve the infection. nonlytic mechanisms slow down the replication kinetics of the virus. This avoids the occurrence of immunopathology and enables the lytic effectors to clear the infection. However, if the cytopathicity of the virus is large relative to its rate of replication (i.e. if $a > (\lambda\beta)^{1/2}$) then both types of immune effectors are beneficial to the host and can, in principle, independently result in resolution of the infection. We have used this theoretical framework to interpret experimental data from mice infected with the noncytopathic LCMV, the more cytopathic influenza virus and the extremely cytopathic VSV. Our discussion has focused on these infections, because aside from nicely reflecting a wide spectrum in terms of viral cytopathicities, these models are well characterized; equally important, the life cycles of the involved viruses allows for rather simple and straightforward virus host relationships. This is critical when first testing the predictions against real life observations. However, our theoretical framework has implications for improving our understanding of the mechanisms required for immunological control in a variety of infections. An interesting example is Hepatitis B Virus (HBV) that appears to be controlled by an intricate balance between a lytic CTL response, nonlytic factors secreted by CTL, and antibody responses [Guidotti and Chisari (1996); Guidotti et al. (1999b)]. CTL-secreted soluble factors have also been reported to 'cure' infected cells from the HBV genome. Other viral infections show more complex life cycles than the ones covered in detail here. For example, certain viruses go through cytopahtic and noncytopathic phases during the course of infection (e.g. Epstein Barr Virus or EBV [Thorley-Lawson and Babcock (1999)]). Other viruses, like HIV, are characterized by varying levels of cytopathicity and replication rates in different cell types (such as macrophages and T cells [van't Wout et al. (1994)]). Although these infections are characterized by more complicated life cycles than assumed in our model, the insights from our theoretical framework can still be applied. They can help us understand how these added complexities influence the viral dynamics as well as the ability of the virus to evade efficient immune-mediated control. For a detailed investigation of such particular infections, the current framework can easily be incorporated into more specific models.

9

Dynamical Interactions between CTL and Antibody Responses

Chapter 8 discussed how lytic and nonlytic CTL responses influence the outcome of infection, what balance of responses is required to control different infections, and how the occurrence of pathology can be avoided. Much of this discussion focused on one type of immune cell, the CTL, that can perform both lytic and nonlytic antiviral activity. As already mentioned in Chapter 8, however, nonlytic CTL responses effect viral spread in the same way as an antibody response. Like nonlytic CTL responses, antibodies also inhibit the rate of virus spread, without killing the infected cells. Antibodies are a major branch of the immune system, and contribute significantly to nonlytic antiviral activity. However, there is an important difference compared to nonlytic CTL responses. Antibodies are produced by B cells that are a separate population of cells from the CTL. Both B cells and CTL proliferate in response to stimulation by the same virus. Therefore, they are in competition with each other for antigenic stimulation [Arnaout and Nowak (2000); Nowak et al. (1995a)]. For example, if the CTL response suppresses virus load to levels that are too low to stimulate the B cells, then a successful B cell/antibody response might not be generated. Conversely, if the B cells reduce virus load to levels that are too low to stimulate the CTL, a successful CTL response will not be established. These competition dynamics add complexity to the situation, because the outcome of competition will influence the relative balance between lytic and nonlytic immune responses that fight a given virus. This balance can in turn determine whether the immune system deals with a pathogen successfully, or whether pathology is observed. This chapter reviews mathematical models that have addressed these interactions. First, a mathematical model is presented that captures the competition dynamics between CTL and antibody responses. We will then apply this model to investigate how the interactions between these two branches of the immune system can influence infection dynamics during acute and chronic phases. For the case of chronic infection, we will explore how the evolution of a persistent virus can shift the balance between lytic and nonlytic immunity over time such that the infection changes from being asymptomatic to being pathogenic. These aspects will be discussed in the context of a specific example that is hepatitis C virus (HCV) infection of humans.

9.1 Modeling Competition between CTL and Antibody Responses

This section presents a mathematical model that includes two different effector responses that fight a viral infection independently: CTL and antibodies. Since it is assumed that both responses rely on antigenic stimulation, the model captures the competition dynamics discussed above [Wodarz (2003)]. The model contains five variables: susceptible host cells x, infected cells y, free virus v, an antibody response w, and a CTL response z. It is given by the following system of ordinary differential equations that describe the change of these populations over time.

$$\dot{x} = \lambda - dx - \beta xv, \tag{9.1}$$
$$\dot{y} = \beta xv - ay - pyz, \tag{9.2}$$
$$\dot{v} = ky - uv - qvw, \tag{9.3}$$
$$\dot{w} = gvw - hw, \tag{9.4}$$
$$\dot{z} = cyz - bz. \tag{9.5}$$

Susceptible host cells are produced at a rate λ, die at a rate dx and become infected by virus at a rate βxv. Infected cells die at a rate ay and are killed by the CTL response at a rate pyz. Free virus is produced by infected cells at a rate ky, decays at a rate uv, and is neutralized by antibodies at a rate qvw. Antibodies develop in response to free virus at a rate gvw and decay at a rate hw. CTL expand in response to viral antigen derived from infected cells at a rate cyz, and decay in the absence of antigenic stimulation at a rate bz.

Infection requires that the basic reproductive ratio of the virus is greater than one. In the absence of an immune responses the system converges to the following equilibrium:

$$x^{(0)} = au/\beta k, y^{(0)} = (\lambda\beta k - dau)/a\beta k,$$
$$v^{(0)} = ky^{(0)}/u, w^{(0)} = 0, z^{(0)} = 0.$$

Now we assume that immune responses can potentially develop. This requires the following conditions: $cy^{(0)} > b$, and $gv^{(0)} > h$. In this case, the following three outcomes can be observed. (Stability conditions have been determined by examining the ability of the immune cell populations to grow from low numbers).

- The CTL response develops and the antibody response cannot become established. This is because the CTL response is strong and reduces virus load to levels that are too low to stimulate the antibody response. It is

9.1 Modeling Competition between CTL and Antibody Responses

described by the following equilibrium. $x^{(1)} = (\lambda uc)/(duc + \beta kb)$, $y^{(1)} = b/c$, $v^{(1)} = ky^{(1)}/u$, $w^{(1)} = 0$, $z^{(1)} = [\beta x(1)v(1) - ay(1)]/py^{(1)}$. This outcome is attained if $gkb/uc < h$ and $c\beta h\lambda/[a(dg + \beta h)] > b$.

- The antibody response develops and a sustained CTL response fails. This is because the antibody response is strong relative to the CTL response and reduces virus load to levels that are too low to stimulate the CTL. This is described by the following equilibrium. $x^{(2)} = \lambda g/(dg + \beta h)$, $y^{(2)} = \beta h\lambda/[a(dg + \beta h)]$, $v^{(2)} = h/g$, $w^{(2)} = ky^{(2)} - uv^{(2)}/qv^{(2)}$, $z^{(2)} = 0$. It is attained if $gkb/uc > h$ and $c\beta h\lambda/[a(dg + \eta bh)] < b$.

- Both CTL and antibody responses develop. This equilibrium is described by $x^{(3)} = \lambda g/(dg + \beta h)$, $y^{(3)} = b/c$, $v^{(3)} = h/g$, $w^{(3)} = ky^{(3)} - uv^{(3)}/qv^{(3)}$, $z^{(3)} = (\beta x^{(3)}v^{(3)} - ay^{(3)})/py^{(3)}$. It is attained if $gkb/uc > h$ and $c\beta h\lambda/[a(dg + \beta h)] > b$.

These outcomes are thus governed by competition between CTL and antibody responses for the virus population. This is because the virus population is a resource that both CTL and antibodies require for survival. Competition can result either in the exclusion of one branch of the immune system, or both branches may coexist. Competition among immune responses has been documented experimentally [Borghans et al. (1999); Freitas and Rocha (2000)], and similar dynamics have been described by [Arnaout and Nowak (2000)].

The following sections will explore how these competition dynamics can influence acute and persistent phases of infection. This will be discussed specifically in the context of hepatitis C virus infection. This is because data indicate that the balance between CTL and antibody responses might determine whether the virus is cleared during acute infection, whether it can establish a persistent, chronic infection, and whether the infection is asymptomatic or pathogenic. HCV primarily infects liver cells. A relatively small percentage of patients clear the virus from the blood, while the rest develop persistent infection that results in liver pathology as long as 10-20 years after infection. The reason for this long asymptomatic phase is not known. Although not completely understood, it is thought that HCV might not be very cytopathic for liver cells, and that liver pathology can be caused by the antiviral CTL response [Chang et al. (1997)] (CTL-induced pathology, see Chapter 7). In addition, clinical data indicate that the virus can escape antibody responses, both during the acute and persistent phases of the infection [Farci (2001); Farci et al. (2000)]. Thus, viral evolution can alter the balance between lytic and nonlytic immunity over time. Therefore, the model will be discussed with HCV in mind, and subsequently, we will discuss biological data in the light of the mathematical insights.

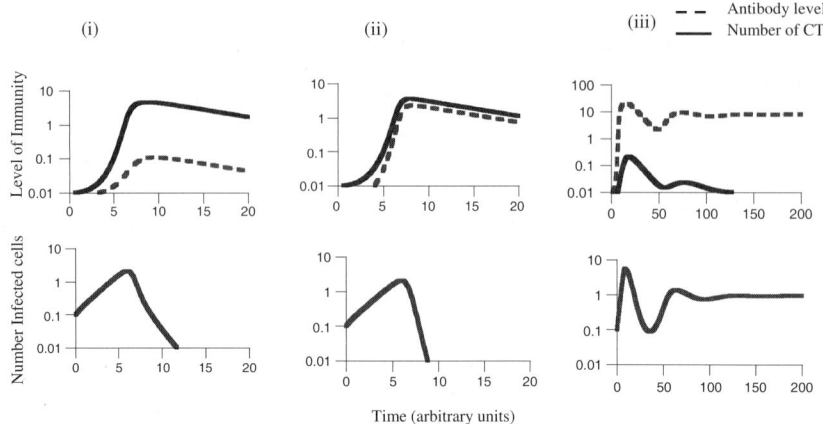

Fig. 9.1. Dynamics during acute infection, according to equations (9.1–9.5). (i) The CTL response is strong relative to the antibody response. Thus, a sustained CTL response develops while the antibody response does not become fully established. The outcome is viral clearance. (ii) Both CTL and antibody responses are sufficiently strong to become fully established. Again, the outcome is virus clearance. (iii) The CTL response is weak relative to the antibody response. The result is persistent infection in the presence of an ongoing antibody response. The CTL response is not sustained. Parameters were chosen as follows: $\lambda = 10$, $d = 0.1$, $\beta = 0.01$, $a = 0.1$, $p = 1$, $k = 1$, $u = 1$, $q = 1$, $h = 0.1$, $b = 0.1$, For (a) $g = 0.5$, $c = 1$. For (b) $g = 1.5$, $c = 1$. For (c) $g = 1$; $c = 0.1$.

9.2 Competition during Acute Infection

We assume that upon infection of the host, the virus population is homogeneous (i.e. there is no antigenic diversity). As the virus population grows, both CTL and antibody responses will start to expand. The outcome of the dynamics in acute infection depends on the relative strengths of the responses. According to the above analysis, three outcomes are possible (Fig 9.1).

(i) The CTL response is dominant relative to the antibody response, and the CTL will clear the infection.

(ii) Both the CTL and antibody responses are equally strong. This is also likely to result in resolution or clearance of infection.

(iii) The CTL response is weak relative to the antibody response. Thus, the antibody response dominates. If we assume that HCV is weakly cytopahtic [Layden et al. (1999); Layden et al. (2000); Neumann et al. (1998)], the antibody response is unlikely to clear the infection. The reason is that while free virus particles are removed, a relatively large pool of infected cells remains because they do not become killed. Hence, the result is persistent infection in the presence of an ongoing antibody response.

Note that it is also possible that the virus is cleared before equilibrium has been reached, because virus load can oscillate before reaching steady state and fall to very low levels. ("dynamic elimination", Chapter 2). In general, simulations indicate that parameter combinations that suppress virus load to low levels at equilibrium also promote the occurrence of "dynamic elimination". Thus, the arguments that are presented in the context of the equilibrium analysis would also apply if the infection is cleared before equilibrium has been reached.

If the virus does persist, the model further suggest that the host will be asymptomatic. The reason is that this scenario is associated with the dominance of an antibody response. We measure the degree of pathology by the number of liver cells. A persistent infection with a weakly cytopathic virus in the presence of an antibody response does not cause significant tissue damage, because there is little killing of virus infected cells.

To summarize, if the virus is assumed to be weakly cytopathic, the model suggests that a strong and sustained CTL response is required to clear the infection. The notion that sustained T cell responses are required for control of HCV infection is strongly supported by experimental and clinical studies [Cooper et al. (1999); Lechner et al. (2000b); Lechner et al. (2000c); Thimme et al. (2001)]. CTL-mediated clearance of infection can also be associated with the presence of antibody responses if these are sufficiently strong. On the other hand, if CTL responses are weak, we can observe persistent infection in the face of a dominant and ongoing antibody response. The model further suggests that this persistent infection is initially asymptomatic.

9.3 Effect of Viral Evolution during Chronic Infection

Here, we explore how this initially asymptomatic persistent infection can change, resulting in the development of liver pathology. We concentrate on the effect of viral evolution towards escape from antibody responses. The following model extends the previous analysis to include viral escape from antibodies.

$$\dot{x} = \lambda - dx - \beta x \sum_{i=1}^{n} v_i, \tag{9.6}$$

$$\dot{y}_i = \beta x v_i - a y_i - p y_i z, \tag{9.7}$$

$$\dot{v}_i = k y_i - u v_i - q v_i w_i, \tag{9.8}$$

$$\dot{w} = g v_i w_i - h w_i, \tag{9.9}$$

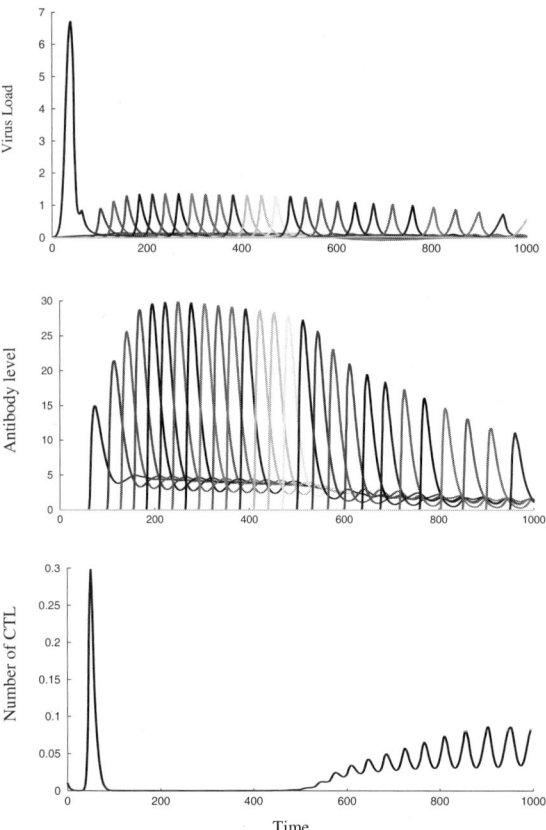

Fig. 9.2. Chronic infection dynamics and viral evolution, according to equations (9.6–9.10). The figure shows the dynamics of the virus population, the antibody response, and the CTL response over time. We assume that the CTL response is weak. Thus, persistent infection develops in the presence of an ongoing antibody response. The presence of antibodies results in the emergence of antibody escape mutants. Each peak of the virus population corresponds to the emergence of a new escape mutant (shown in different shades). The antibody response adapts to these new variants by creating new specificities. As the virus population evolves towards increased diversity, the weak CTL response expands. This coincides with a reduction in antibody responses. Once the CTL response becomes more dominant, liver pathology can set in. Parameters were chosen as follows: $\lambda = 1$, $d = 0.1$, $\beta = 0.03$, $a = 0.1$, $p = 1$, $k = 2.5$, $u = 2$, $q = 1$, $g = 2$, $h = 0.1$, $c = 0.1$, $b = 0.2$.

$$\dot{z} = cz \sum_{i=1}^{n} y_i - bz. \tag{9.10}$$

The model assumes that n virus variants can be generated that vary in the antibody epitopes. Antibody escape has mostly been studied in the context of the hypervariable regrion 1 (HVR1), but the model description applies to any mutation that confers resistance to antibodies. For simplicity we assume that the variants only differ in their antibody epitopes; otherwise, they are identical (for example, they replicate at the same rate). The results do not, however, depend on this simplification. Viruses of strain i are denoted by v_i, and cells infected with strain i are denoted by $y_i (i = 1..n)$. Each virus strain can elicit an antibody response that is specific for this strain w_i. For simplicity we assume that the strength of the antibody response against the individual variants is identical. While the antibody response against the different variants are likely to be characterized by different efficacies, the dynamics in question are not changed by this simplification. We also include a CTL response in the model. We assume that the CTL response is crossreactive and can recognize all antibody escape variants. For the present arguments we do not need to consider antigenic escape from CTL responses in the model.

In the following, we investigate how virus evolution influences the dynamics between HCV and the immune responses from acute infection through the chronic phase, assuming that the CTL response is weak. This is done by a combination of analytic and numeric methods. As outlined above, a weak CTL response can result in persistent infection, and this leads to the dominance of antibody responses (Fig 9.2). Because of persistent replication, variants evolve that escape the antibodies. These new variants, in turn, elicit new antibody responses, and in this way, increased antigenic diversity develops over time (Fig 9.2). The outcome of infection as a function of the number of virus strains n is given by the following equilibria.

$$x^{(4)} = \frac{\lambda g}{(dg + n\beta h)}, y_i^{(4)} = \frac{\beta\lambda h}{a(dg + n\beta h)},$$

$$v_i^{(4)} = h/g, w_i^{(4)} = \frac{ky_i^{(4)} - uv_i^{(4)}}{qv_i^{(4)}}, z^{(4)} = 0.$$

Thus, as the virus population evolves, virus load slightly increases due to the accumulation of antigenic variants, and the antibody response broadens as a result of this diversification. The overall number of liver cells is, however, predicted to remain constant, which corresponds to absence of pathology (Fig 9.3). These dynamics change if the number of antigenic variants crosses a threshold given by $n > badg/([\beta h(c\lambda - ba)]$. If this condition is fulfilled, the CTL response can expand and increase in dominance relative to the antibody response (Fig 9.2). Thus, while the CTL responsiveness is weak, the accumulation of mutants that escape from the antibodies increases antigenic drive, and this promotes increased expansion of the weak CTL. However, this CTL response contributes little to virus control. Instead, it marks the beginning of liver pathology (Fig 9.3): CTL kill the infected cells and can thus contribute to tissue damage [Chang et al. (1997)]. Invasion of the CTL response after accumulation of antigenic diversity is described by the following equilibrium expressions.

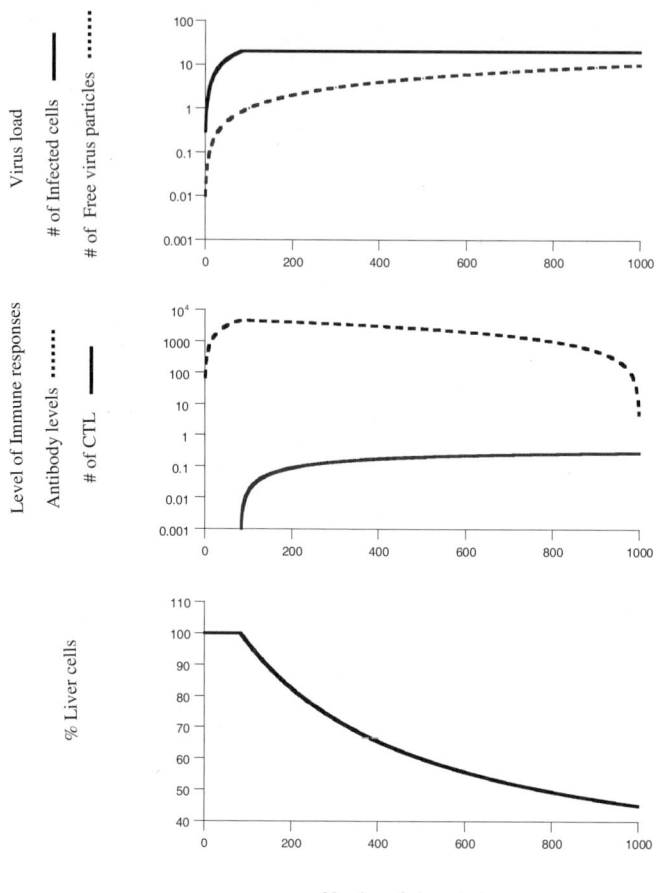

Fig. 9.3. Outcome of chronic infection as a function of the number of antibody escape mutants n, according to equations (9.6–9.10). Equilibrium values of virus load, immune responses, and the percentage of liver cells are plotted. Accumulation of virus diversity allows the CTL response to expand. The CTL can induce liver damage. As the number of antibody escape mutants increases, liver damage becomes stronger. This can correspond to disease progression and to an increase in liver damage over time. While the equilibrium values give us a good idea of how the outcome of infection depends on the number of antibody escape variants, the actual population sizes can be different if the escape dynamics are sufficiently fast so that the equilibrium will not be visible (see Fig 9.2). The general trend, however, remains robust. Parameters were chosen as follows: $\lambda = 10$; $d = 0.1$, $\beta = 0.03$, $a = 0.1$, $p = 1$, $k = 2.5$, $u = 5$, $q = 1$, $g = 10$, $h = 0.1$, $c = 0.01$, $b = 0.2$.

$$x^{(5)} = \frac{\lambda g}{dg + n\beta h}, y_i^{(5)} = b/nc, v_i^{(5)} = h/g,$$

$$w_i^{(5)} = \frac{(kgb - nuhc)}{ncqh}, z^{(5)} = \frac{n\beta h(\lambda c - ab) - dgab}{[pb(dg + n\beta h)]}.$$

The degree of pathology that results from CTL invasion is a function of the number of virus strains (Fig 9.3). Continued escape from antibody responses shifts the balance between antibodies and CTL further towards the CTL. Since antibodies prevent, and CTL induce tissue pathology, the amount of liver damage increases with the accumulation of antibody escape variants. When the number of antigenic variants crosses a threshold, the CTL response attains maximum dominance relative to the antibody response. This threshold is given by $n > gkb/uch$. At this point, CTL-induced pathology is expected to be at its maximum (Fig 9.3). Furthermore, as this threshold is attained, virus evolution is expected to stop, because target cell limitation does not allow invasion of additional virus variants in the face of significant liver destruction. Note in Figs 9.2 & 9.3 that there is no significant change in virus load as the dynamics progress from the asymptomatic phase towards liver destruction. This suggests that the development of disease will not correlate significantly with levels of virus load.

9.4 Application: Experimental Data on HCV Infection

The modeling framework helps us understand and interpret experimental data on immune responses against HCV. The role of CTL and antibodies for the resolution of HCV infection is debated in the literature [Farci et al. (2000); Klenerman et al. (2000)]. Studies of the acute phase of the infection showed that both humans or chimpanzees who cleared the virus from blood developed strong and sustained CTL responses [Chang et al. (2001); Cooper et al. (1999); Lechner et al. (2000a); Lechner et al. (2000b); Lechner et al. (2000c); Thimme et al. (2001)]. In chimpanzee studies, CTL-mediated clearance is associated with the absence of strong antibody responses [Cooper et al. (1999)], and this supports the notion of competition between the two branches of the immune system. Humans and chimpanzees who developed persistent infection were characterized by an initial CTL response that was not sustained at high levels beyond the acute phase of infection [Cooper et al. (1999); Lechner et al. (2000a); Lechner et al. (2000b); Lechner et al. (2000c); Thimme et al. (2001)]. Persistent infection has, however, been observed to be associated with vigorous antibody responses [Farci et al. (2000); Major et al. (1999)], again pointing to the occurrence of competition. Thus, it was argued that a strong CTL response is crucial for the resolution of infection. The theoretical results presented here agree with this conclusion. On the other hand, it has been argued that antibody responses are crucial for deciding the outcome of infection [Farci et al. (2000)]. Farci et al. demonstrated that patients who resolved the infection showed evolutionary stasis in the virus population. Patients who developed chronic infection showed accumulation of genetic variation in antibody epitopes. It was concluded that escape from antibody responses allows the virus to persist. The importance of viral evolution for persistence has also been stressed by others [Forns et al. (1999); Weiner et al. (1992)]. The theory presented here argues that antigenic escape might not be the reason for viral persistence, but the consequence of viral persistence. Continued productive infection in the presence of an immune response is the most efficient scenario for the evolution of antigenic escape [Wodarz and Nowak (2000a)]. This argument does not, however, diminish the importance of antibodies in HCV infection. If a persistent infection is established, the model suggests that antibodies are crucial for keeping the patient asymptomatic, and that escape from antibodies allows the balance of immune responses to be shifted in favor of weak CTL that cause liver pathology. Thus, the importance of different types of immune responses should be considered in a broader setting than just in the context of clearing the virus during acute infection. The relative balance of antibody and CTL responses can be crucial for deciding whether the persistent infection remains asymptomatic, or whether pathology is observed.

A central concept in this analysis is that viral evolution towards escape from antibodies might drive disease progression. This concept is very difficult to test with data. Studies have quantified sequence diversity and the rate of virus evolution in patients that differ in the severity of liver disease [Cabot et al. (2000); Curran et al. (2002); Farci (2001); Lyra et al. (2002); Major et al. (1999); Thomson et al. (2001)]. Even if virus evolution does drive disease progression - as suggested here - this does not mean that we expect higher virus diversity or faster evolution in patients with severe compared to mild disease. The amount of viral diversity and the rate of evolution is a complex function of the degree of immune-mediated viral suppression, virus load, and the number of susceptible host cells [Wodarz and Nowak (2000a)]. If virus load is suppressed efficiently, reduced replication does not allow a fast accumulation of mutations. On the other hand, immune-mediated suppression of the virus allows for the coexistence of different antigenic variants [Regoes et al. (1998)]. If immune responses are diminished, however, competition between virus strains becomes an important factor that can lead to a reduction in viral diversity [Nowak et al. (1991)]. Moreover, as the number of susceptible cells becomes limiting - for example because of liver destruction - virus evolution may slow down or stop, because new antigenic variants cannot invade [Regoes et al. (1998)]. Therefore, if pathology develops as a result of virus evolution towards antigenic escape, it is possible that patients with severe disease show reduced levels of virus diversification and evolution compared to patients with milder disease. Data on virus diversity have given rise to different and contradictory results [Cabot et al. (2000); Curran et al. (2002); Farci (2001); Lyra et al. (2002); Major et al. (1999); Thomson et al. (2001)]. Studies with HCV infected patients report the emergence of antibody escape mutants early after infection [Farci et al. (2000)], while data from the chimpanzee model argue against the frequent occurrence of antibody escape mutants during the early stages of the infection [Major et al. (1999)]. While the chimpanzee data argue against a role of escape for virus persistence, this study does not address the

9.4 Application: Experimental Data on HCV Infection 135

role of escape for disease progression, which occurs over a much longer period of time. Curran et al. [Curran et al. (2002)] reported a consistent accumulation of amino acid changing substitutions in patients with mild liver disease. In patients with severe liver disease, however, they reported significantly lower rates of virus evolution. The observation that the virus continuously evolves in patients with mild disease supports the notion that ongoing virus evolution can eventually shift the dynamics towards pathology. A reduced rate of evolution in patients with severe disease is also consistent with the theoretical arguments presented here. As severe pathology develops, the number of susceptible host cells is reduced and this can prevent invasion of further antigenic variants. The hypothesis that viral evolution can drive disease progression is further supported by the observation that the virus diversity, as well as the ratio of amino acid changing to silent substitutions was higher in individuals with severe liver disease [Cabot et al. (2000); Curran et al. (2002)]. Similar patterns have been observed in studies of viral evolution following liver transplants in patients with mild and severe disease [Lyra et al. (2002)]. Investigation of the expected patterns of diversity and rates of evolution in patients with different severity of disease under various assumptions will benefit from further mathematical modeling.

Another factor that can influence patterns of viral evolution is the ability of the specific immune responses to adapt to the changing virus population. The model assumes that new escape variants can always trigger new antibody specificities. Therefore, the virus continuously evolves away from the immune response. Clinical data from HCV infected patients suggest that specific CD4 T helper cell responses are impaired [Lechner et al. (2000c)]. Helper cell impairment can in turn compromise the ability of the neutralizing antibody response to adapt to new virus variants [Klenerman et al. (2000)]. If this happens, virus evolution towards increased antigenic diversity in antibody epitopes is expected to stop. It is not clear for how long the antibody response can continue to adapt to the virus population. As the disease progresses, the antibody response is more likely to fail to adapt to the virus population. This could also contribute to the reduced rate of evolution seen in HCV infected patients with severe liver disease.

It is an interesting observation that accumulation of antigenic diversity and the development of disease does not correlate with a significant increase in virus load in the model. Upon development of symptoms, virus load may be relatively small: While CTL induce pathology in the model, they can suppress viremia to a certain degree at the same time. Variations in virus load observed between patients is thus unlikely to be due to differences in the amount of antigenic diversity and disease status, but to differences in other host parameters. Hence, the models suggests that there is no obvious correlation between pathology and virus load. This is consistent with clinical data [de Araujo et al. (2002); Manzin et al. (1997); Pontisso et al. (1999); Puoti et al. (1999)].

9.5 Summary

This chapter expanded the consideration of lytic and nonlytic effector mechanisms. In contrast to Chapter 8, where the same CTL were assumed to have both lytic and nonlytic antiviral activity, we now assumed that lytic and nonlytic effector mechanisms derive from separate populations of immune cells: CTL and antibodies. Both branches require antigenic stimulation to mount a response, and they are therefore in competition. The balance between lytic and nonlytic effector activity can therefore depend on the outcome of this competition. If CTL exclude antibodies, then the balance can be shifted towards lytic responses. If antibodies exclude CTL, then the balance can be shifted towards nonlytic immunity. If both branches coexist, the balance is more even. Finally, we discussed how the evolution of antigenic escape by the virus can shift the balance of immune responses over time, and how this can shift the outcome of the dynamics from lack of disease towards pathology. HCV infection could be an example of where the competition between CTL and antibodies might determine the course of infection, and where escape from antibodies during chronic infection could shift the dynamics from an asymptomatic to a pathogenic state.

10
Effector Molecules and CTL Homeostasis

Chapters 7, 8, and 9 have discussed effector molecules in relation to their direct antiviral activity: to kill infected cells or to inhibit the spread of the virus from cell to cell by nonlytic means. We discussed the role of the different types of effector mechanisms for the successful resolution of infection, and how the exact balance between lytic and nonlytic effector mechanisms can determine whether pathology is observed or not. However, experimental data indicate that the effector molecules might have a broader function than just antiviral activity [Stepp et al. (2000)]. In addition, they might serve as regulatory molecules that determine the extent to which the population of CTL expands in response to antigenic stimulation. Experimental data indicate that the amount of effector molecules that are secreted by CTL can influence the peak size of the CTL response, as well as the number of CTL found in the memory phase after the infection has been resolved. Of particular interest in this respect are mice that have deficiencies in either perforin or IFN-γ. Perforin is the main effector molecule that is responsible for CTL-mediated lysis, while IFN-γ is known to be a CTL-secreted soluble mediator that inhibits viral replication in several infections, for example LCMV. A deficiency in either perforin or IFN-γ, or both, has been observed to result in higher levels of CTL both in the acute and the post acute phase of infections [Bartholdy et al. (2000); Matloubian et al. (1999); Stepp et al. (2000)].

Matloubian et al [Matloubian et al. (1999)] reported that perforin might be an important regulatory molecule that determines CTL numbers during the response against LCMV infection. Perforin deficient mice were characterized by persistent infection, higher levels of CTL, a reduced decline of activated CTL, and by pathology that resulted in death of the animals. Mortality was completely reversed by depleting the CTL *in vivo*. Thus it was hypothesized that perforin downregulates the CTL response during chronic infection, and therefore prevents excessive CTL-mediated killing that can result in tissue damage. Bartholdy et al [Bartholdy et al. (2000)] investigated the dynamics of LCMV infection in IFN-γ deficient mice. They considered the relatively slowly replicating Armstrong strain. While wild-type mice cleared the infection, a persistent infection was established in IFN-γ deficient mice. Again, the persistent infection was accompanied by an ongoing CTL response that was sustained at higher levels compared to wild-type animals. Furthermore, a higher proportion of the CTL had an activated phenotype and were actively cycling compared to wild-type animals. However, no pathology was observed in this experiment. Instead, the persisting virus was controlled by the elevated CTL response in the long-term at low loads. Badovinac et al [Badovinac et al. (2002); Badovinac et al. (2000)] investigated the dynamics of the bacterium *Listeria monocytogenes* in perforin- and IFN-γ deficient mice. They also found that the absence of these effector molecules correlated with an increased expansion and a decreased contraction phase of CTL, which resulted in elevated numbers of CTL in the acute and the post acute phase of the infection. In addition, they found that the immunodominance hierarchy was altered in knockout animals. Based on their experiments, they argued that expansion of the CTL is controlled by perforin, while IFN-γ controlled the death phase of the CTL and the immunodominance hierarchy.

The reasons for these observations have been debated in the literature. Mathematical models have played an important role in this respect [Bartholdy et al. (2000); Thomsen et al. (2000); Wodarz (2001b)]. They can suggest mechanisms that underly the experimental observations, and they can be used to test whether certain hypotheses are consistent with experimental observations or not. This chapter starts with a possible explanation of the experimental data, as suggested by mathematical modeling. We then review arguments that counter this hypothesis. Finally, we discuss further mathematical and experimental work that has addressed this controversy.

10.1 CTL Homeostasis and Predator–Prey Dynamics

The observation that the absence of effector molecules results in elevated levels of CTL can be explained without invoking a separate regulatory role of the effector molecules [Wodarz (2001b)]. Instead, this observation could come about by the predator–prey type dynamics that underly the interactions between CTL responses and infectious agents. The infection can be considered the prey that becomes killed. The CTL can be considered the predator that reproduces in the presence of the prey (virus). Consider the ecological scenario where foxes are predators and rabbits are prey. If each fox is inefficient at capturing rabbits, many rabbits are found. Because many rabbits are around, many foxes also persist as a consequence. On the other hand, if each fox is very efficient at capturing rabbits, then few rabbits remain in the system. Fewer rabbits can in turn support fewer foxes. Effector molecules, such as perforin and IFN-γ, determine the efficiency with which CTL can "capture" and and remove pathogens. If effector molecules are present at normal levels, then the CTL are efficient at reducing virus load. Low virus load can only support few CTL because CTL proliferation depends on antigenic stimulation. On

10.1 CTL Homeostasis and Predator–Prey Dynamics

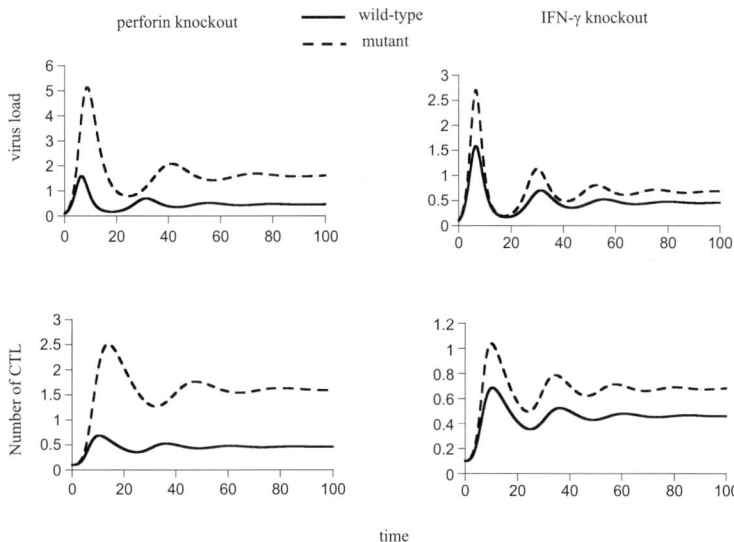

Fig. 10.1. CTL dynamics in perforin deficient and IFN-γ deficient hosts, as predicted by the mathematical model (8.1–8.3). Absence of perforin and IFN-γ results in elevated virus load and elevated levels of CTL as a result of the increased antigenic drive. Parameters were chosen as follows: $\lambda = 10$, $d = 0.1$, $\beta = 0.01$, $a = 0.2$, $c = 0.1$, $b = 0.1$; for wild-type $p = q = 1$; for mutant $p = q = 0.1$. Note there is still some effector activity left in the mutant scenario since there can be other sources of lytic and nonlytic CTL-mediated activity.

the other hand, if the effector molecules are absent, then the CTL are less efficient at reducing virus load. Higher virus loads can in turn maintain a larger number of CTL. This can be shown by computer simulations (Fig 10.1) that are based on the mathematical model that includes virus dynamics as well as the presence of lytic (perforin) and nonlytic (IFN-γ) CTL responses (8.1–8.3). The activity of perforin is expressed in the parameter p. Relatively high values of p correspond to wild-type animals, while reduced values of p or $p = 0$ corresponds to perforin knockout mice. The activity of IFN-γ is expressed in the paramter q. Relatively high values of q correspond to wild-type animals, while knockout animals are characterized by reduced values of q or $q = 0$. Fig 10.1 compares the wild-type scenario to the perforin- and IFN-γ knockout simulations. In both cases, the virus persists at higher levels. This results in higher peak numbers of CTL during the acute phase of the infection, and the persistently replicating virus drives continuous CTL expansion that leads to higher numbers of CTL during the memory phase of the infection.

The simulations can be taken further to investigate the conditions under which the knockout of effector molecules also results in CTL-induced tissue pathology (see Chapter 7). Consider IFN-γ deficiency first. According to the model, the degree of CTL induced pathology observed in the absence of IFN-γ depends on the rate of viral replication (see Chapter 7). If the virus replicates slowly, there is minimal target cell depletion and the host is likely to survive with the virus and CTL coexisting. This corresponds to the experiments with LCMV Armstrong infection in IFN-γ deficient mice described above [Bartholdy et al. (2000)]. If the virus replicates with a high rate, on the other hand, strong depletion of target cells is predicted to occur in the absence of IFN-γ, and the host is likely to die. Such an outcome has been observed with the faster replicating LCMV Traub strain [Bartholdy et al. (2000)]. Similar considerations apply to the absence of perforin. Even if perforin is completely absent, the host can still potentially die from CTL-induced pathology in the context of a fast replicating virus. This is because FAS-mediated lysis, a less efficient and perforin-independent mechanism to kill infected cells, can still cause pathology. This explains why some perforin deficient hosts not only show elevated levels of CTL, but also tissue pathology [Matloubian et al. (1999)].

10.2 Effector Molecules and Immunodominance

In experiments with the intracellular bacterium *Listeria monocytogenes*, it was found that the absence of effector molecules not only elevates the overall level of the antiviral CTL, but that it might also alter the immunodominance hierarchy of different CTL clones that react to different epitopes [Badovinac et al. (2000)]. In particular, IFN-γ knockout mice showed this effect. On the other hand, perforin knockout mice did not show an altered immunodominance hierarchy.

The consideration of immunodominance brings additional complexity to the theoretical arguments, because we must now consider multiple CTL clones that react against the same virus, but against different epitopes. A given CTL clone is said to be immunodominant if it reaches higher abundances relative to other CTL clones. Mathematical models that consider multiple CTL clones that are directed against different viral epitopes suggest that immunodominance is the result of competition of CTL for antigenic stimulation by the virus [Nowak et al. (1995a); Nowak et al. (1995b)] (Chapter 5). Each CTL clone requires antigenic stimulation to proliferate. CTL clone 1 may reduce virus load to levels that are too low to effectively stimulate CTL clone 2. Consequently, CTL clone 1 will be present in significantly higher abundances than CTL clone 2. CTL clone 1 is said to be immunodominant. CTL clone 2 can even go extinct. According to mathematical models, the competitive ability of CTL is mostly determined by the rate at which they respond to antigen and proliferate. Thus, for the presence of IFN-γ to alter the immunodominanc hierarchy, IFN-γ must be able to influence the rate of CTL proliferation. For instance, it has been documented that IFN-γ can upregulate MHC expression and thus promote the induction of CTL responses [Heise et al. (1998a); Heise

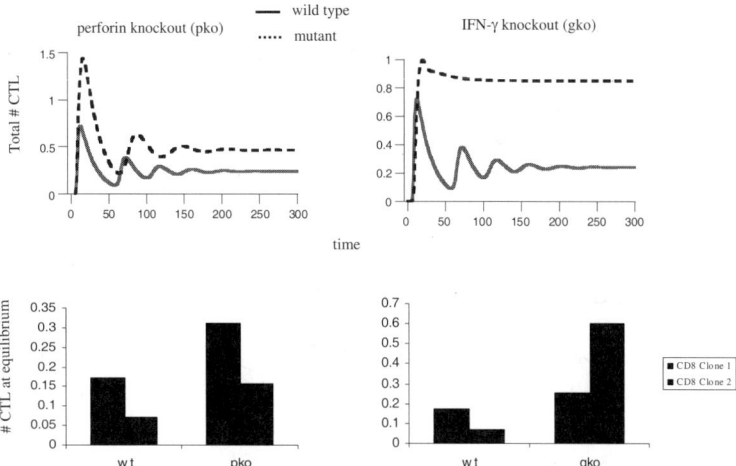

Fig. 10.2. Immunodominance in perforin deficient (pko) and IFN-γ deficient (gko) hosts, as predicted by the mathematical model (10.1–10.4). Absence of perforin results in a higher level of specific CTL, but not in an altered death phase or immunodominance. Absence of IFN-γ does result in a reduced death phase of the CTL and in altered immunodominance. Note that immunodominance hierarchies are shown at equilibrium in this figure in order to keep the graphs clear and concise. However, the immunodominance hierarchies in the model are already established right at the beginning of the infectious process when CTL responses start to expand. Parameters were chosen as follows: $\lambda = 10$, $d = 0.1$, $\beta = 0.02$, $a = 0.1$, $p_1 = 1$, $p_2 = 0.1$, $q_1 = 30$, $q_2 = 5$, $c_1 = 0.01$, $c_2 = 0.02$, $b = 0.05$, $\epsilon = 10$, $f = 0.001$. For pko: $p_1 = 0.2$, $p_2 = 0.02$. For gko: $q_1 = 0.5$, $q_2 = 0.5$.

et al. (1998b)]. The following equations present a model that describes two CTL clones that are directed against different epitopes of the same virus. The CTL are assumed to be able to lyse infected cells by perforin, and also secrete IFN-γ, which can inhibit viral replication, and which can promote CTL proliferation.

$$\dot{x} = \lambda - dx - \frac{\beta x y}{(q_1 z_1 + q_2 z_2) + 1}, \tag{10.1}$$

$$\dot{y} = \frac{\beta x y}{(q_1 z_1 + q_2 z_2 + 1)} - ay - y(p_1 z_1 + p_2 z_2), \tag{10.2}$$

$$\dot{z}_1 = \frac{(c_1 + f q_1) y z_1}{\epsilon z_1 + 1} - b z_1, \tag{10.3}$$

$$\dot{z}_2 = \frac{(c_2 + f q_2) y z_2}{\epsilon z_2 + 1} - b z_2. \tag{10.4}$$

x denotes the quantity of uninfected host cells, y is the quantity of infected host cells, z_1 is the quantity of specific CTL clone 1, and z_2 is the quantity of specific CTL clone 2. As in the basic model for virus infection (2.2–2.4), uninfected cells are produced with a rate λ and die with a rate d. They become infected by the pathogen at a rate β and die at a rate a. CTL lyse infected cells at a rate p_i. They also secrete IFN-γ, which inhibits microbial replication at a rate q_i. Two factors determine the responsiveness of the CTL: the basic expansion rate of the cells c_i and the aforementioned secretion of IFN-γ, which enhances the rate of CTL proliferation at a rate fq_i (e.g. through MHC upregulation by IFN-γ). The rate of CTL proliferation is assumed to saturate at high CTL abundances, as determined by the parameter ϵ (See Chapter 2). Finally, CTL die with a rate b in the absence of antigenic stimulation. Knockout of perforin and IFN-γ is represented in the model by a reduction in the parameters p_i and q_i respectively.

Assume that CTL clone 1 has a lower basic rate of proliferation in response to antigenic stimulation but secretes larger amounts of IFN-γ than CTL clone 2. Because of the larger amounts of IFN-γ secreted by CTL clone 1, the proliferation of this clone can be sufficiently enhanced such that it is the dominant response (Fig 10.2). If the production of IFN-γ is reduced in both CTL clones by the same amount (equivalent to a knockout), it is possible that clone 2 becomes the dominant response (Fig 10.2). This is because in this IFN-γ deficient situation, IFN-γ does not contribute significantly to increasing the proliferation rate of the CTL. Since CTL clone 2 has the higher basic rate of proliferation in response to antigenic stimulation, it will dominate. In contrast to this result, a deficiency of perforin does not result in altered levels of immunodominance in the model (Fig 10.2).

Therefore, the experimental observations in perforin and IFN-γ knockout mice can be explained in relatively simple terms. The elevated levels of CTL during the acute and post acute phases of infection can come about through the simple predator–prey dynamics that underly the interactions between a pathogen and CTL responses, without the need to invoke separate regulatory roles. A change in the immunodominance hierarchy requires in addition that the effector molecules influence the rate of CTL expansion upon antigenic stimulation. In particular, they need to enhance CTL proliferation in response to antigenic stimulation. This has been well documented in the literature in the context of IFN-γ [Heise et al. (1998a); Heise et al. (1998b)], but not for perforin. This explains the altered immunodominance hierarchy in IFN-γ deficient, but not in perforin deficient mice.

10.3 The Role of Antigen for CTL Proliferation

The arguments presented in the previous section rely on the fact that CTL respond to antigenic stimulation: less effector activity results in higher virus load; higher virus load correlates with a larger antigenic drive that leads to a larger number of CTL. Recently, however, experimental data have revealed that the development of CTL responses occurs by a process called *programmed proliferation* (see Chapter 2) [Kaech and Ahmed (2001); Mercado et al. (2000); van Stipdonk et al. (2003); van Stipdonk et al. (2001)]. According to this concept, a single encounter with antigen is necessary to induce CTL expansion and differentiation. The encounter with antigen activates the naive CTL. Subsequently, the CTL undergo a program of proliferation that includes approximately 7-10 division events. The population of CTL differentiates into effector cells, declines to a certain degree, and then settles around a memory level. The processes of division, differentiation, contraction, and memory generation are not influenced by further exposure to antigen. That is, they occur even if the CTL never see antigen again. Further antigenic stimulation is only required if the generated memory CTL need to be become reactivated. In this case, another program of expansion and differentiation may be triggered.

If CTL expansion occurs according to a program that is independent of antigenic stimulation, it can be argued that the level of CTL, or the immunodominance hierarchy, cannot be driven by virus load or competition for antigen. Thus, elevated levels of CTL cannot be explained by increased antigenic drive that is a consequence of reduced effector activity. Instead, the increased levels of CTL in perforin and IFN-γ knockout animals (as well as the altered immunodominance hierarchy in IFN-γ mice) requires that the effector molecules have separate regulatory functions that determine CTL homeostasis [Badovinac et al. (2002); Badovinac et al. (2000); Stepp et al. (2000)]. In the next section we discuss mathematical models and experimental data that address this controversy.

10.4 Programmed CTL Proliferation and the Role of CTL Effectors

Here, we consider a model that takes into account explicitly the concept of programmed CTL expansion. Assuming that perforin and IFN-γ only perform effector function and do not have any separate regulatory activity, we will investigate whether and how a deficiency of these effector molecules influences the level of the CTL responses. We consider the model that describes programmed CTL proliferation (2.18–2.23), assuming that CTL can have both lytic and nonlytic effector activity.

According to this model, the absence of effector molecules is predicted to have no effect on the number of antiviral CTL if the CTL only undergo a single round of proliferation. That is, naive cells undergo approximately 7-10 programmed divisions, differentiate into effectors, and eventually into memory cells. In other words, the memory cells do not become restimulated again. This, however, might only occur in a restricted set of circumstances. It is quite likely that after only one round of programmed proliferation some of the virus, or some antigen derived from the virus, is still present in the host. In this case, the memory CTL will become restimulated and will undergo another round or programmed proliferation. If this occurs, the model predicts that the lack of effector molecules will lead to elevated levels of antiviral CTL. This is because after the first round of CTL proliferation, reduced effector activity will leave more virus in the system, and this results in a stronger stimulation of the memory CTL. In fact, in the context of a persistent infection, the properties of the program model converge to those of the simplified predator prey equation that assumes that CTL division requires continuous antigenic stimulation. This has been demonstrated in Chapter 2. Consequently, the mechanism that was derived from the simple predator–prey like models to account for the elevated level of CTL in perforin and IFN-γ knockout mice remains robust in the context of programmed proliferation as long as the memory CTL become restimulated.

10.5 Effector Molecules and CTL Homeostasis in VSV Infection

Vesicular stomatitis virus (VSV) has some special properties that allow us to investigate further, whether reduced effector activity alone can account for the elevation in the number of antiviral CTL (without invoking separate regulatory functions of the effector molecules) [Christensen et al. (2004)]. While CTL responses develop against VSV, they do not contribute to virus clearance. VSV infection is very virulent in mice. It either kills the host, or it is cleared by an antibody response. Antibody mediated clearance occurs before significant levels of antiviral CTL can be observed. Nevertheless, the VSV specific CTL expand and differentiate into effector and memory CTL. This is probably because they become stimulated while the virus is still in the system, and thus the antigen independent proliferation program is triggered. By the time effector activity is achieved, however, the virus is already cleared. Therefore, the effector molecules do not perform antiviral effector activity and do not influence virus load. Hence, if the reduced antiviral activity by itself is responsible for the elevation of the number of specific CTL, then we would expect that with VSV infection of mice, perforin or IFN-γ do not influence the level of CTL. If, on the other hand, perforin and IFN-γ have separate regulatory activity which determines CTL homeostasis, then we should also observe elevated numbers of CTL in knockout mice infected with VSV.

Unlike what has been observed when perforin and IFN-γ are essential for pathogen clearance, neither molecule was found to play an important role in regulating the kinetics of the VSV specific CTL response, which is characterized by the absence of antiviral activity *in vivo* [Christensen et al. (2004)]. Therefore, in this setting, it appears that the effector molecules do not have separate regulatory activity. This is consistent with the argument that in the other experiments discussed in this chapter, the elevation of specific CTL numbers in knockout mice is brought about by the reduced effector activity and the resulting higher antigenic drive itself.

10.6 Summary

This chapter has discussed how mathematical modeling has been useful in suggesting a mechanism that could account for the elevated levels of antiviral CTL in perforin- and IFN-γ knockout mice, and for the altered immunodominance in IFN-γ deficient animals. The models suggest that the basic predator–prey like interactions that underly the dynamics between replicating pathogens and CTL responses could lead to the experimental observations: reduced effector activity leads to higher antigenic drive that in turn leads to a higher number of CTL. It is thus not necessary to invoke separate regulatory roles for these effector molecules. Mathematical models further suggest that this mechanism is even consistent with the concept of an antigen independent CTL proliferation program, as long as the memory CTL become restimulated after the first round of CTL expansion. Experiments with VSV infected mice support these notions. In this scenario, CTL effector molecules do not contribute to fighting the virus and do not influence virus load. The absence of the effector molecules does not result in significantly altered CTL levels in this case. While all these results show that the experimental data are consistent with the most parsimonious explanation, they do not exclude the possibilities that these effector molecules have additional functions that should be investigated further. An understanding of these mechanisms is also important from a medical point of view. The disease called "familial hemaphagocytic lymphohistiocytosis" or FHL is characterized by unregulated and self destructive CTL responses [Stepp et al. (2000)]. Patients react to simple infections with inflammatory processes characterized by the activation of T cells and macrophages. There is evidence that FHL patients harbor genetic and functional defects of the effector molecule perforin [Stepp et al. (2000)].

11
Virus-Induced Subversion of CTL Responses

So far, this book has discussed how CTL responses can fight viral infections, leading to their successful resolution and clearance. On the other hand, viruses have the ability to avoid clearance by a variety of mechanisms. A very prominent mechanism that has been documented in the context of several pathogens is antigenic escape. Viral epitopes acquire mutations that prevent the CTL response from recognizing the epitope. Consequently, the infected cell is not attacked by the CTL. HIV is probably the best known virus that shows extensive antigenic escape [McMichael et al. (1996); McMichael and Phillips (1997); McMichael and Rowland-Jones (2001); Phillips et al. (1991); Price et al. (1997a); Price et al. (1999)]. Because HIV has a relatively high mutation rate, escape mutants are readily generated and this can contribute to the inability of the CTL response to fight HIV effectively, and it might contribute to the eventual development of AIDS. HCV infection is another example of a human pathogen that can readily acquire mutations, allowing it to escape from immune responses [Farci et al. (2000)]. Another mechanism to avoid immune-mediated clearance is to establish a latent infection [Thorley-Lawson and Babcock (1999)]. This means that cells can be infected by the virus, but once inside the cell, the virus is silent and does not produce further progeny viruses for prolonged periods of time. If the virus is silent, the CTL cannot recognize that the cell is infected, because no viral proteins are produced. If the cell is sufficiently long-lived, it can carry the virus and provide a reservoir for viral persistence. At certain time intervals, the virus can reactivate in the cells and start producing new virus particles. This is called the lytic phase of the infection, and the CTL response tends to prevent growth of the virus to high numbers. Thus, while the CTL manage to control the virus in this case, latent infection of cells prevents clearance of the virus.

These are examples of viral strategies to escape a fully functional CTL response. An alternative strategy is to impair the functionality of the CTL response itself (Fig 11.1). This leads to an ineffective response that is unable to deal with the infection, even if the virus does not escape from the response by any other means. In other words, the virus actively attacks the immune

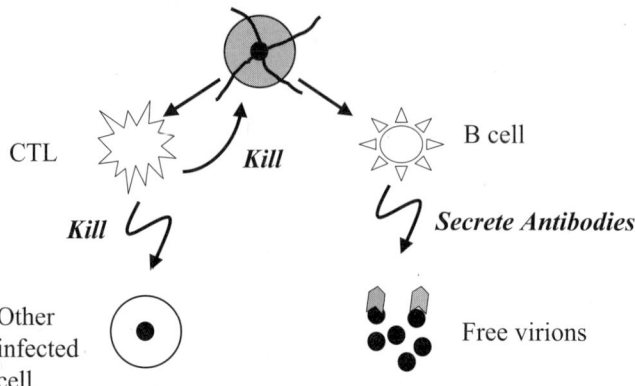

Fig. 11.1. Virus-induced suppression of the CD4 T helper cell response. The most prominent example is HIV infection. Helper cells are the central component of the immune system that orchestrate specific immune effector response such as CTL and antibodies. If T helper cells are infected by the virus, they can become killed and this can compromise the ability of the immune system to mount effective responses. The helper cells can be killed either by the virus itself, or by CTL that kill the infected T helper cells.

system. Active, virus-induced impairment of the CTL response is the subject of this chapter. A major mechanism by which this can be achieved is the infection and killing of CD4 T helper cells Fig (11.1). This population of cells is of central importance in the generation of any antiviral effector responses. If CD4 T cell help is impaired, the host fails to mount effective CTL or antibody responses against the pathogen [Borrow et al. (1996); Borrow et al. (1998); Thomsen et al. (1996); Thomsen et al. (1998)].

Recently, experiments have generated more detailed insights into the role of CD4 T cell help for the development of CTL responses. Experiments suggest that initial CTL expansion and differentiation in the acute phase does not depend on help [Borrow et al. (1996); Borrow et al. (1998); Janssen et al. (2003); Shedlock and Shen (2003); Sun and Bevan (2003); Sun et al. (2004); Thomsen et al. (1996); Thomsen et al. (1998)]. The maintenance of memory and/or the reactivation of the memory cells, however, crucially depends on help. The mechanism that underlies this observation is unclear. Some studies suggested that the helper cells somehow "program" the CTL during acute infection and that the subsequent availability of help has no influence [Shedlock and Shen (2003); Sun and Bevan (2003)]. On the other hand, a recent study indicates that the presence of help may be continuously required to maintain memory in the long-term [Sun et al. (2004)]. In any case, the absence of help might lead to a response that is initially normal during the acute phase of infection, but then quickly loses its effectiveness because the restimulation of memory CTL is impaired. As discussed extensively in Chapter 3, the absence of effective and sustained CTL memory responses can be a reason for the failure to clear a virus, and for the establishment of persistent infection.

As described in Chapter 4, "help" can be delivered to CTL via antigen presenting cells (APCs), most prominently dendritic cells [Ridge et al. (1998); Schoenberger et al. (1998)]. That is, the CD4 T helper cells directly interact with the APCs, which results in the activation of the APCs. Activated APCs, in turn, directly interact with CTL, which promotes the proliferation and differentiation of the CTL. Thus, the easiest way for a virus to impair CD4 cell help is to infect and kill either the CD4 T helper cells directly, or to infect and kill the APCs. Both options apply to HIV, which has been shown to impair the T helper cell response against itself significantly [Douek et al. (2002); Rosenberg et al. (1999)]. LCMV infection is also thought to impair helper cell responses. LCMV infects a variety of cell types that include APCs and CD4 T helper cells [Moskophidis et al. (1995a)]. While LCMV is noncytopathic, the initial acute CTL response that develops against the virus might kill infected APCs and helper cells and thereby remove the very stimulus that it needs to be sustained. HCV is another example where the virus specific helper cell response is impaired, although the exact mechanism of this impairment is unclear. Even if the virus does not kill APCs and helper cells, it can disable the responses by other mechanisms. An example is anergy, where the the presence of too much virus might result in the failure of the response to develop.

This chapter reviews mathematical models that describe the dynamics between the CTL response and a virus that infects and impairs its specific CD4 T helper cell responses. We investigate what outcomes of infection are expected to be observed, and how host an viral parameters determine whether the CTL can resolve the infection or not. Theory will be applied to specific case studies such as LCMV infection of mice, and SIV/HIV infection.

11.1 A Basic Model for Virus-Induced Impairment of Help

As mentioned above, help influences the ability of memory cells to expand. In the absence of help, memory cells cannot be restimulated efficiently, and CTL responses cannot be sustained beyond the acute phase of infection [Janssen et al. (2003); Shedlock and Shen (2003); Sun and Bevan (2003); Sun et al. (2004)]. In order to investigate the properties of helper cell impairment mathematically, we therefore modify a basic mathematical model that distinguishes between two subpopulation of CTL: the memory CTL that are CTL precursors (CTLp) without effector activity; and the effector CTL that have antiviral activity (2.12–2.13). In this model, the CTLp proliferate in response to antigenic stimulation and subsequently differentiate into effector cells. To include the effect of help, we assume that the rate of CTLp proliferation is not only proportional to antigen, but also to the number of cells that can deliver helper signals for the CTL [Wodarz et al. (1998); Wodarz and Nowak (1999); Wodarz et al. (2000c)]. To take into account the ability of the virus to impair help, we further assume that these helper cells are targets for infection by the virus, and can therefore be killed by infection. Denoting the susceptible population of helper cells by x, the population of infected cells by y, the population of CTLp by w, and the population of CTL effectors by z, the model is given by the following set of ordinary differential equations.

$$\dot{x} = \lambda - dx - \beta xy, \qquad (11.1)$$
$$\dot{y} = \beta xy - ay - pyz, \qquad (11.2)$$
$$\dot{w} = cwxy - qwy - b_1 w, \qquad (11.3)$$
$$\dot{z} = qwy - b_2 z. \qquad (11.4)$$

The rate of CTLp expansion is given by $cwxy$ and is therefore proportional to both antigen y and the number of helper cells x. The proliferating CTLp differentiate into effectors with a rate qwy. The CTLp die with a rate $b_1 w$ and the effector CTL die with a rate $b_2 z$. Note that this is a simplified model. As mentioned above, the initial CTL expansion in acute infection is not dependent on help. Help is only required for the stimulation of memory CTL. This model does not include an initial helper independent response, but assumes that CTL expansion always depends on help. This is because we are interested in the long-term dynamics and equilibrium outcomes, and not in the details of the acute infection dynamics. Thus, numerical simulations of the acute phase are expected to be inaccurate. In particular, the initial amount of CTL expansion in the absence of help will be lower than expected *in vivo*. Our analysis, however, does not concentrate on this phase.

Because this model is based on the general virus dynamics equations 2.2–2.3), establishment of infection requires that the basic reproductive ratio of the virus is greater than one (see Chapter 2). If the virus establishes an infection, two alternative outcomes can be observed. First, the CTL response can go extinct and the virus persists at relatively high levels. This is described by

$$x^{(1)} = a/\beta,$$
$$y^{(1)} = \lambda/a - d/\beta,$$
$$w^{(1)} = 0,$$
$$z^{(1)} = 0.$$

In this case, the virus wins the fight. It successfully impairs the helper cell response such that the virus can replicate unopposed.

Alternatively, a sustained CTL response is established that controls the virus at relatively low levels over the long-term. The exact virus load will depend on the strength of the CTL response. If virus load is sufficiently low, it will correspond to virus extinction in practical terms. This outcome is described by the following equilibrium expressions.

$$y^{(2)} = \frac{b_1}{cx^{(2)} - q},$$

$$w^{(2)} = \frac{b_2(cx^{(2)} - q)(\beta x^{(2)} - a)}{b_1 qp},$$

$$z^{(2)} = \frac{\beta x^{(2)} - a}{p},$$

$$x^{(2)} = \frac{(\lambda c + qd - b_1 \beta) + \sqrt{(\lambda c + qd - b_1 \beta)^2 - 4dc\lambda q}}{2dc}.$$

In this case, the CTL response wins. It maintains the upper hand, and controls the virus in the long-term.

In summary, we observe two outcomes: Failure to contain the virus and successful control of the virus. Note that failure to control does not necessarily correlate with the complete extinction of the CTL response in reality. The absence of help might result in the maintenance of inefficient CTL that persist in the absence of help and that do not contribute significantly to virus control. Such a population of inefficient "helper independent" CTL can be included in the model, as discussed below. The following sections examine the conditions under which loss of control is observed, and when the CTL response is successfully established.

11.2 What Determines the Outcome of Infection?

Here, we examine under which conditions the CTL manage to control the virus, and when the CTL fail and go extinct in the model. The parameter region of the model can be divided into two regions, according to the following condition:

$$(\frac{\lambda}{l} Komatsuetal.] - \frac{d}{\beta})(c\frac{l}{K} omatsuetal.]\beta - q) > b_1.$$

The most important parameter that determines the outcome of this condition is the replication rate of the virus β relative to the strength of the CTL response c. If the viral replication rate is relatively low, the condition holds. If the viral replication rate lies above a threshold, the condition is violated. If the inequality is true, the CTL response will persist. However, if the inequality is not satisfied, the situation is more complicated. In this case, one has to distinguish between two parameter regions that mostly depend on the rate of viral replication.

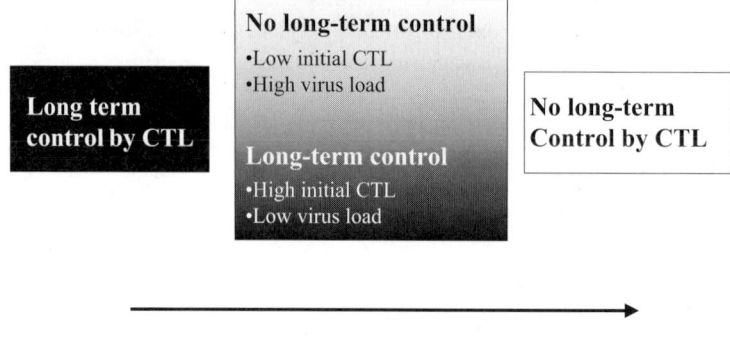

Fig. 11.2. Outcomes of the mathematical model (11.1–11.4). There are two outcomes. Either, the CTL response goes extinct and the virus replicates at high levels, or a sustained CTL response become established and we observe long-term control of the infection. Which outcome is achieved depends on the rate of viral replication relative to the CTL responsiveness of the host, and can depend on the initial conditions. If the viral replication rate lies below a threshold, the outcome is CTL extinction. If the viral replication rate lies above a threshold, the outcome is long-term CTL-mediated control of the virus. If the viral replication rate is intermediate, both outcomes are stable and the outcome depends on the initial conditions. Low initial virus load and a high initial number of CTL promote the CTL control outcome, while high initial virus load and low initial numbers of CTL promote the CTL extinction outcome.

- If the rate of viral replication is intermediate and lies below a threshold, then both the CTL persistence and the CTL exhaustion equilibria are stable. In other words, we observe bistability. Now the outcome of the system depends on the initial conditions. A small initial virus load promotes CTL persistence, while a high initial virus load promotes extinction of the CTL response. The initial number of CD4 T helper cells is also significant with a low initial CD4 T helper cell count driving the system towards CTL extinction. The initial number of CTL is also important. Low initial numbers of CTL (i.e. a naive host) promotes CTL extinction, while a high initial number of CTL promotes CTL-mediated virus control. This outcome is observed if the CTL control equilibrium expressions are real ($[\lambda c + qd - b_1\beta]^2 > 4dc\lambda q$) and positive ($x^{(2)} > q/c$ or $x^{(2)} > a/\beta$).

- If the rate of viral replication is higher and lies above a threshold, then the CTL control equilibrium becomes unstable and the only outcome is CTL extinction. This is observed if the CTL control equilibrium expressions are complex or negative.

In summary, the rate of viral replication is a major determinant of the outcome of infection. If it is relatively low and lies below a threshold, the only possible outcome is CTL-mediated control of the infection. If the viral replication rate is faster, we enter a parameter region where both the CTL control outcome and the CTL extinction outcome are stable. In this case, the outcome of infection depends on the initial conditions. If the rate of viral replication crosses a final threshold, then the only possible outcome is the extinction of the CTL response.

11.3 Robustness of Predictions

We have discussed the dynamics of virus-induced helper cell impairment in the context of a specific model that took account of CTL precursors and CTL effectors. As pointed out in Chapter 2, uncertainties remain regarding the expansion and differentiation pathway of CTL upon antigenic stimulation, and about the exact mechanism by which viruses impair CD4 T cell help. The results discussed here do, however, not depend on these uncertainties. The same properties are observed in a much simpler and more general mathematical model that does not make specific assumptions about these details [Komarova et al. (2003)]. It is summarized as follows.

We consider a model that contains two variables: the virus population y and a population of immune cells, e.g. CTL, z. We assume that the degree of immune expansion depends on virus load, and that the response inhibits virus growth. The model is given by the following pair of differential equations:

$$\dot{y} = y g_r(y) - yz, \qquad (11.5)$$
$$\dot{z} = z f(y). \qquad (11.6)$$

The virus population grows at a rate that is described by the function $g_r(y)$. This is a function that depends on the amount of virus y and on the parameter r, denoting the viral replication rate. The virus population becomes inhibited by the immune response at a rate yz. Immune expansion is determined by virus load y and is described by the function $f(y)$. The generic shape of functions $g_r(y)$ and $f(y)$ is presented in Fig 11.3. Positive values of these functions indicate growth, and negative values correspond to decay in the population. Consider the virus growth function $g_r(y)$. On the trivial side we make the assumption that the higher the replication rate of the virus r the higher the viral growth rate. The only other assumption is that virus growth is density dependent: growth slows down at higher virus loads; when virus load crosses a threshold, growth stops and the virus population declines. This corresponds to target cell limitation where the virus runs out of cells to infect.

Now we turn to the function describing immune expansion, $f(y)$. We make the assumption that the presence of the virus can both stimulate and impair immunity, depending on virus load y. If virus load lies below a threshold, the

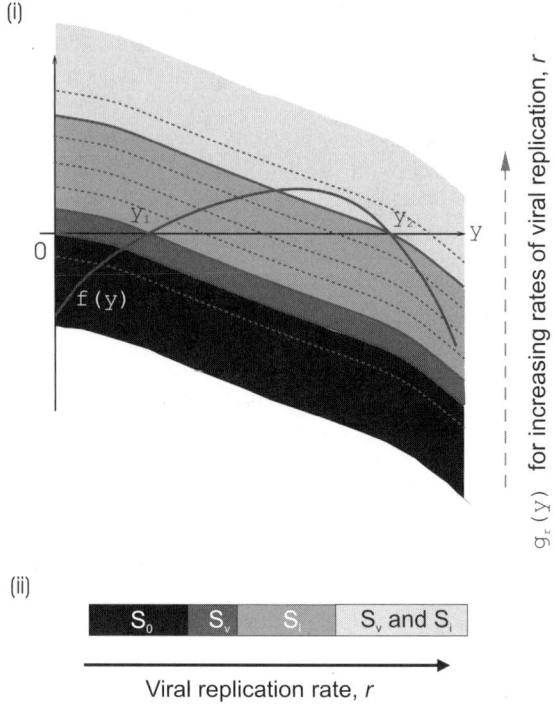

Fig. 11.3. Diagram explaining the behavior of the general mathematical model (11.5–11.6). The model is defined by the virus growth function $g_r(y)$ and the immune expansion function $f(y)$. As explained in the text, the function $g_r(y)$ depends on the viral replication rate, r, while the function $f(y)$ does not. **(i)** The functions $f(y)$ and $g_r(y)$ are plotted versus y. $f(y)$ is the humped curve labeled in the figure. The rest of the curves represent $g(r)$ for different valus of r. The function $f(y)$ has two fixed roots, y_1 and y_2. The parameter r determines at which value of $y = y_*$ the function $g_r(y)$ equals zero. If the value of r lies above a threshold so that $g_r(y) = 0$ for $y > y_2$, we are in the bistable parameter region. If the value of r is smaller so that $g_r(y) = 0$ for $y_1 < y < y_2$, the only stable outcome is immune control S_i. If the value of r is still smaller so that $g_r(y) = 0$ for $0 < y < y_1$, only the virus equilibrium S_v is stable. Finally, if the value of r is very small, then $g_r(y) = 0$ for $y < 0$ and the infection cannot become established in the first place, S_0. **(ii)** This is a schematic diagram that summarizes how the outcome of infection depends on the viral replication rate, r, as detailed above.

rate of immune expansion is negative. That is, levels of antigen are too low to induce a response. If virus load lies above a threshold, the rate of immune expansion is also negative because immune impairment outweighs antigenic stimulation. Thus, high virus loads inhibit immunity. Immune expansion is positive for intermediate virus loads, because antigenic stimulation is strong relative to immune impairment. The precise conditions $f(y)$ and $g_r(y)$ have to

11.3 Robustness of Predictions

satisfy for the results to hold true are given in [Komarova et al. (2003)]. This framework makes minimal assumptions and defines a collection of models. Hence, results derived from this framework are robust and do not depend on the particular form of the equations. We analyze the behavior of the whole class of models. We find three outcomes.

- In the trivial case S_0 there is no infection and no immune response.
- Alternatively, the virus can establish an infection in the absence of an immune response, S_v. We will refer to the equilibrium S_v as the *virus equilibrium* because the virus can grow unchecked and cause pathology directly or indirectly.
- Finally, the virus can establish an infection that is controlled by an immune response, S_i. We refer to this outcome as the *immune control equilibrium*.

As in the more detailed model (11.1–11.4), the outcome depends on the replication rate of the virus, parameter r. This is described as follows.

(i) If the replication rate is very small, the virus cannot infect the host and the system converges to S_0.

(ii) If the replication rate of the virus crosses a threshold, an infection can be established, but the amount of antigenic stimulation is too low to trigger sustained immunity. The system converges to S_v.

(iii) If the replication rate is higher and crosses another threshold, levels of antigen are sufficient to trigger sustained immunity. The system converges to the equilibrium describing long-term immunological control, S_i.

(iv) If the viral replication rate is still higher and crosses a final threshold, the immune response can be significantly impaired. In this parameter region, both the immune control (S_i) and the virus equilibrium (S_v) are stable, and the outcome of infection depends on the initial conditions in the same way as discussed in the more detailed model (11.1–11.4).

Therefore, the behavior of this much simpler, and more generalized model is very similar to that of the more detailed and mechanistic model (11.1–11.4). However, there is one difference: in model (11.1–11.4) a very high rate of viral replication results in the instability of the CTL control outcome and the dynamics always converge to the CTL extinction equilibrium. This does not happen in the simplified model (11.5–11.6). This difference is, however, not very meaningful. Although in the simplified model, the CTL control outcome does not become stable for high rates of viral replication, the trajectories are less and less likely to lead to the CTL control outcome the faster the viral replication rate is. Therefore, the CTL control outcome will for practical purposes never be attained at high rates of viral replication.

This model will be considered in more detail in Chapter 12, because it forms the basis for exploring how therapy can be used to restore CTL-mediated control in immunosuppressive infections.

11.4 Experimental Verification: CTL Exhaustion in LCMV Infection

If adult mice are infected with LCMV, a variety of outcomes can be observed (reviewed in Chapter 1). Two of these outcomes are CTL mediated clearance and CTL exhaustion [Moskophidis et al. (1993a)]. CTL exhaustion means that after the primary response, the CTL go extinct and the virus replicates persistently at high levels. Because LCMV is noncytopathic, the mice survive despite persistent infection at high virus loads. Therefore, similar to the mathematical model, we observe a CTL control and a CTL extinction outcome. Experiments have determined when each outcome is observed in LCMV infected mice. In accordance with the mathematical model, it has been found that the outcome of LCMV infection depends on the cell tropism of LCMV, the replication rate of LCMV, and the initial conditions.

A central feature of our mathematical model is that the virus actively impairs CD4 T cell help, and that this can be the reason for CTL extinction. Different LCMV strains are characterized by different cell tropisms, i.e. they infect different cell types [Ahmed and Oldstone (1988)]. For example, the Armstrong strain is predominantly neurotropic. On the other hand, the C13 and Docile strains are lymphotropic, i.e. they infect immune cells. Among the immune cells, they predominantly infect CD4 T helper cells and antigen presenting cells such as macrophages. Data indicate that CTL exhaustion correlates with the ability of LCMV to infect such immune cells [Ahmed et al. (1988); Borrow et al. (1995); King et al. (1990); Oldstone et al. (1988); Villarete et al. (1994)]. This supports the mathematical notion that CTL extinction is the result of active virus-mediated immune impairment.

The main viral determinant of whether the CTL control the infection or go extinct in the mathematical model is the replication rate of the virus. Faster virus replication promotes CTL extinction in the model. This also holds true in LCMV infection of mice [Christensen et al. (2001); Moskophidis et al. (1995a); Moskophidis et al. (1993c); Thomsen et al. (2000)]. Different strains replicate at different rates. Faster replicating strains, such as Docile and Traub, are more likely to induce CTL exhaustion than slowly replicating strains, such as Armstrong. In addition, the lack of interferon-mediated suppression of viral replication promotes the occurrence of CTL exhaustion [Christensen et al. (1999); van den Broek et al. (1995a); van den Broek et al. (1995b)]. Further, the outcome can depend on the route of infection [Moskophidis et al. (1995a)]. Subcutaneous infection is less efficient at causing CTL exhaustion than intravenous infection. Subcutaneous infection might slow down the initial spread of the virus considerably, compared to intravenous infection. The importance of viral replication for the outcome of infection is also demonstrated by experimental data in Fig 11.4.

Finally, the model suggests that the outcome of infection can depend on the initial conditions [Moskophidis et al. (1995a); Moskophidis et al. (1993c); Wodarz et al. (1998)]. This is confirmed by experimental data (Fig 11.4). A

11.4 Experimental Verification: CTL Exhaustion in LCMV Infection

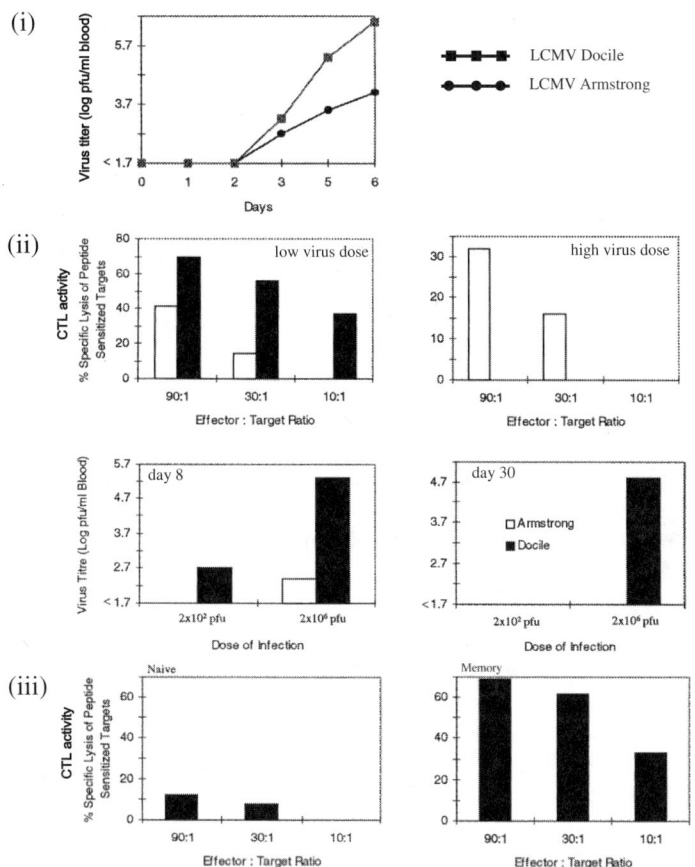

Fig. 11.4. Experimental data on CTL exhaustion in LCMV infection taken from [Wodarz et al. (1998)] (i) In vivo growth rate of LCMV docile (faster replicating) and LCMV Armstrong (slower replicating). (ii) CTL persistence versus CTL exhaustion, depending on the initial virus load. With low initial virus load ($2x10^2$ pfu), the CTL response against both strains persists with higher lelves of lysis observed for LCMV Docile infected mice. Virus titers are kept at low or undetectable levels. With high dose infection ($2x10^6$ pfu), the CTL response against LCMV Armstrong persists while it is exhausted with LCMV Docile infection. Accordingly, virus titers in LCMV Docile infected mice persist at relatively high levels. (iii) Naive (primary) challenge versus memory rechallenge. The difference is that in memory mice the initial number of CTL is elevated. CTL-mediated activity remained much higher in memory compared to naive mice.

higher initial virus load correlates with a higher efficiency of exhausting the CTL response. The faster the replication rate of a strain, the lower the initial virus dose necessary to induce a decline of the specific CTL. The dependence of the course of infection on the initial virus load is shown in Fig 11.4. While

LCMV Docile fails to induce exhaustion of the CTL response at an initial dose of $2x10^2$ pfu, the specific CTL vanished when mice were infected with $2x10^6$ pfu. In accordance with the model, the initial numbers of T cells also plays an important role. As shown in Fig 11.4, CTL exhaustion is less likely to occur in mice that have been immunized and have a high number of memory CTL at the time of infection, compared to mice that are naive and have low numbers of CTL at the time of infection. In addition, data indicate that CD4 T cell depletion promotes the occurrence of CTL exhaustion.

11.5 Helper-Dependent versus Independent CTL Responses

The basic model of helper cell impairment (11.1–11.4) was characterized by two outcomes: CTL control and CTL extinction. In LCMV infected mice, these two outcomes are observed, and experiments support the model predictions regarding the conditions under which CTL extinction occurs. CTL extinction is, however, not a very prominent observation among viral infections. Consider HIV infection. There are also two outcomes that correspond to control versus lack of control. Some patients, called long-term nonprogressors, control the infection at very low virus loads for more than 15 or 20 years. These patients are characterized by high levels of CTL that are thought to contribute to this state [Harrer et al. (1996a); Harrer et al. (1996b); Rosenberg et al. (1997); Rosenberg et al. (1999)]. On the other hand, most patients are typical progressors that develop AIDS within a variable period of time (on average 5-10 years). These patients have relatively high virus loads, and weak CTL responses. In this case, the CTL fail to control the infection. Yet, the virus specific CTL do not go extinct (although they might go extinct at the very end of the disease process). How can this be reconciled with the basic model of virus-induced immune impairment (11.1–11.4)?

The CTL included in the basic model of immune impairment (11.1–11.4) can be thought of as so called *helper dependent CTL*. That is, their maintenance is strictly dependent on the presence of CD4 T cell help. In addition, they are fully functional and can differentiate into memory cells that can be reactivated repeatedly by the persistent infection. If help is limiting, this population of CTL can go extinct. Apart from this, we also consider a population of so called *helper independent CTL*. These are CTL that develop in the absence of help. They do not differentiate into a full memory phenotype and are therefore less efficient at fighting the virus infection. It is possible that such inferior CTL are maintained in the absence of specific T cell help because cytokines secreted by a variety of immune cell types could induce limited proliferation. These CTL are, however, unlikely to control an infection. Rather they are maintained as a result of continuous virus persistence and a high activation state of the immune system. Therefore, rather than to distinguish between the presence and absence of specific CTL, we distinguish between

11.5 Helper-Dependent versus Independent CTL Responses

two types of CTL responses that are qualitatively different: a helper dependent response that can control an infection in the long-term, and a helper independent response that is maintained by persisting antigenic drive and cannot control the infection in the long-term. In the following, we modify the basic model of immune impairment (11.1–11.4) to include such a population of helper independent CTL.

Virus dynamics are described by the basic model of virus replication (2.2–2.4). We add two types of CTL responses to this model: (i) The helper independent response is captured in the variable z_1 that represents effector cells. The CTL expand at a rate $c_1 y z_1$ and decay at a rate $b_1 z_1$. They kill infected cells at a rate $p_1 y z_1$. Obviously, CTL expansion does not depend on help. Consequently, a population of memory cells are not included, since memory generation requires help. (ii) The helper-dependent response is modeled in a similar way as in the basic impairment model (11.1–11.4). That is, we distinguish between memory CTLp w and effector CTL z_2. This is because in the presence of help, memory responses develop that can be restimulated by further exposure to antigen. The memory CTL can proliferate in the presence of help and antigen with a rate $c_2 xyw$ and decay in the absence of antigenic stimulation with a rate $b_2 w_2$. Th memory CTL differentiate into effector CTL with a rate $c_2 qyw$, and effector cells die with a rate hz_2. Effector CTL kill infected cells with a rate $p_2 y z_2$. The model is given by the following set of differential equations.

$$\dot{x} = \lambda - dx - \beta xy, \tag{11.7}$$

$$\dot{y} = \beta xy - ay - p_1 z_1 y - p_2 z_2 y, \tag{11.8}$$

$$\dot{z}_1 = c_1 z_1 y - b_1 z_1, \tag{11.9}$$

$$\dot{w} = c_2 xyw - c_2 qyw - b_2 w, \tag{11.10}$$

$$\dot{z}_2 = c_2 qyw - hz_2. \tag{11.11}$$

A basic difference between helper-dependent and helper-independent CTL in the model concerns the life span of the response at low levels or in the absence of antigen. We assume that the helper dependent response is long-lived in the absence of antigen, manifested in the memory CTLp population. On the other hand, the helper-independent response is relatively short lived in the absence of antigen, because it is assumed that memory CTL are not generated in this case. Hence, $1/b_2 \gg 1/b_1$.

We assume that the basic reproductive ratio of the virus is greater than unity. That is, the virus can initially grow in the host. In this primary phase of the infection, both helper dependent and independent CTL responses can in principle expand. The system subsequently settles to one of two alternative equilibria. The first outcome (Fig 11.5i) describes establishment of the helper-dependent CTL response that can be maintained at low levels of antigen. This response can control the virus infection in the long-term. The

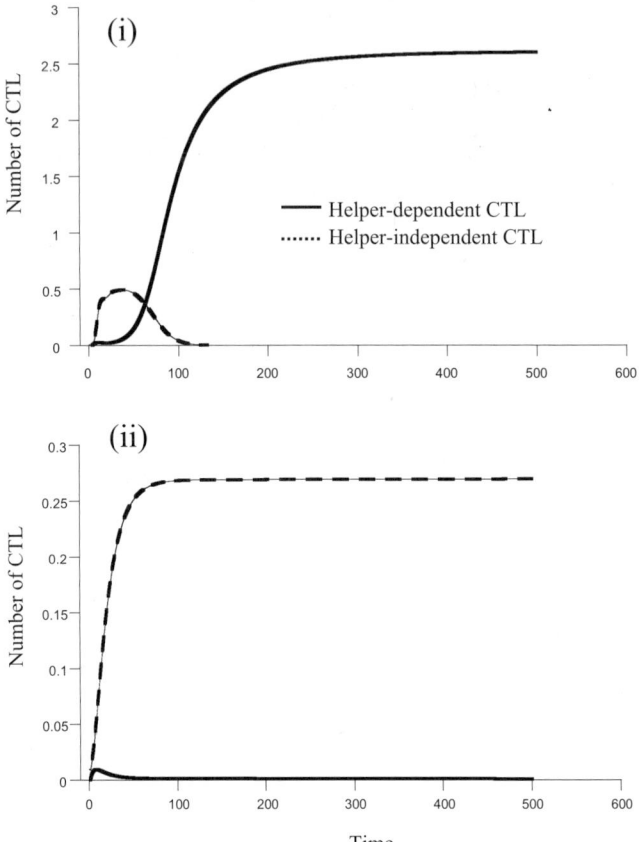

Fig. 11.5. The establishment of helper-dependent and helper-independent CTL responses in the model (11.7–11.11). (i) If the helper-independent CTL response is strong relative to the replication rate of the virus, then virus spread is limited. Significant amounts of immune impairment are avoided and a more efficient helper dependent memory CTL response becomes established. The helper-independent CTL response goes extinct. (ii) If the helper-independent CTL responsiveness is low relative to the rate of viral replication, then virus spread is fast. This results in significant amounts of helper cell impairment and the failure to develop a helper-dependent CTL response. Instead, the helper-independent response persists, maintained by relatively high levels of antigenic stimulation. Parameters were chosen as follows: $\lambda = 1$, $d = 0.1$, $a = 0.2$, $p_1 = 1$, $p_2 = 1$, $b_1 = 0.1$, $c_2 = 0.05$, $b_2 = 0.01$, $q = 0.5$, $h = 0.1$, For (a) $c_1 = 0.1$, $\beta = 0.3$. For (b) $c_1 = 0.05$, $\beta = 0.8$.

helper-independent response goes extinct. This is described by the following equilibrium.

$$x^{(1)} = \frac{\lambda}{d + \beta y^{(1)}},$$

11.5 Helper-Dependent versus Independent CTL Responses

$$z_1^{(1)} = 0,$$
$$w^{(1)} = \frac{h z_2^{(1)}}{c_2 q y^{(1)}},$$
$$z_2^{(1)} = \frac{y^{(1)}(c_2 \beta q - c_2 a) + b_2 \beta}{c_2 p_2 y^{(1)}},$$

where

$$y^{(1)} = \frac{c_2(\lambda - dq) - b_2 \beta - \sqrt{[c_2(\lambda - dq) - b_2 \beta]^2 - 4\beta c_2 q d b_2}}{2\beta c_2 q}.$$

The second outcome (Fig 11.5ii) describes the failure of the helper-dependent CTL response to become established. In this case, the virus can replicate persistently at significant levels, and this virus replication maintains a helper-independent CTL response that can suppress virus load to a certain degree. This is described by the following equilibrium.

$$x^{(2)} = \frac{\lambda c_1}{d c_1 + b_1 \beta},$$
$$y^{(2)} = b_1/c_1,$$
$$z_1^{(2)} = \frac{\beta x^{(2)} - a}{p_1},$$
$$w^{(2)} = 0,$$
$$z^{(2)} = 0.$$

The behavior of the model is similar to the simpler model described above (11.1–11.4). The helper-dependent response always becomes established if

$$\beta < \frac{c_1 [c_2 b_1 (\lambda - qd) - b_2 c_1 d]}{b_1 (c_2 b_1 q + b_2 c_1)}. \tag{11.12}$$

This condition can be simplified by making two assumptions: (i) the helper dependent CTL response is long-lived, i.e. the value of b_2 is small. (ii) the rate of target cell generation is significantly faster than the rate of target cell death, i.e. $\lambda \gg qd$. In this case the above condition can be written as $\beta < c_1 \lambda / b_1 q$. Hence establishment of a helper-dependent CTL response is promoted by a slow rate of viral replication relative to the responsiveness of the helper-independent CTL. This is because the extent of early viral replication determines the amount of virus-induced immune impairment. The stronger the helper-independent CTL response and the slower the rate of viral replication, the slower the initial spread of the virus population. This paves the way for the rise of a helper dependent memory CTL response required for long-term control of the infection.

If the above condition (11.12) is not fulfilled, then we have to distinguish between two parameter regions. If the replication rate of the virus β lies below a threshold relative to the strength of the helper-independent CTL response, then two outcomes are stable (this threshold could not be defined mathematically): the establishment of the helper-dependent CTL, and the establishment of the helper-independent CTL. Which outcome is observed depends on the initial conditions. Establishment of a helper-dependent CTL response is promoted by a low initial virus load, a high initial number of helper-independent CTL, and a high initial CD4 cell response. On the other hand, if the replication rate of the virus β lies above a threshold, then the helper-dependent CTL response cannot become established and the only possible outcome is lack of virus control in the presence of helper-independent CTL.

Therefore, the conditions which decide whether a helper dependent CTL response controls the infection, or whether there is lack of control in conjunction with the maintenance of helper-independent CTL are qualitatively similar to the conditions which decide whether we observe CTL extinction or CTL-mediated virus control in the basic immune impairment model (11.1–11.4). In both cases, faster virus replication promotes lack of long-term control, and we observe a parameter region in which the outcome of infection depends on the initial conditions.

In addition, an important result from this extended model (11.7–11.11) is that the presence of a helper independent CTL response promotes the establishment of a helper-dependent response. The stronger the helper-independent response, the higher the chances that a helper-dependent memory response becomes established. In addition, the presence of a high initial number of helper-independent CTL promotes a switch to the helper-dependent memory CTL response. This finding is interesting to consider in the light of competition dynamics; especially in the light of previous theoretical studies that analyze the dynamics of multiple CTL clones directed against the same virus population [De Boer and Perelson (1994); Nowak (1996)]. In these models, different populations of CTL compete for the same resource, i.e. the virus population. The CTL clone characterized by the highest responsiveness outcompetes all other CTL clones (see Chapter 5). This is because it reduces virus load to a level too low to stimulate the competitively inferior CTL populations. In the model analyzed here (11.7–11.11), there are two types of CTL that grow in response to the same virus population, and this also results in competition dynamics. The presence of a helper-dependent CTL response leads to the extinction of the helper-independent response, since it can reduce virus load to very low levels. However, the reverse is not true. In the contrary, the presence of a helper-independent response promotes the establishment of a helper-dependent response. This is because the helper-independent CTL response prevents the occurrence of very high virus loads, and thus prevents strong levels of helper cell impairment. Hence, competition is asymmetric: while the helper-dependent CTL response can exclude the helper-independent response, the helper-independent response lowers immune impairment and promotes the establishment of the helper-dependent response.

11.6 Immune Impairment and the Level of Immune Responses

A central argument throughout this chapter was that virus-induced immune impairment results in significantly reduced levels of CD4 T cell help, and this can lead lack of virus control due to the extinction of the virus specific CTL, or the maintenance of inefficient CTL responses. This is also shown in clinical data that measure the HIV specific CD4 T cell responses in typical HIV infected patients and in long-term nonprogressors who do not develop AIDS after more than 15-20 years [Rosenberg et al. (2000); Rosenberg et al. (1997); Rosenberg et al. (1999); Rosenberg and Walker (1998)]. long-term nonprogressors have significantly higher levels of HIV specific CD4 T cell responses compared to typical patients. However, this simple and intuitive reasoning can be misleading. It is important to recognize that the *level* of the response does not necessarily correlate with the *strength* of the response [Betts et al. (2001); Wodarz et al. (2000a); Wodarz et al. (2001)]. This can be demonstrated with a mathematical model.

So far, we have considered a simplified model that included a general population of helper cells that could be infected. Emphasis was placed mostly on the virus specific CTL response. Here, we focus on the virus specific CD4 T helper cell response. That is, we distinguish between two subpopulations of CD4 T helper cells: the virus specific population of helper cells, and the general population of helper cells. We further assume that virus specific helper cells need to be activated by the virus in order to become infected. This has been documented with several infections, for example HIV and HTLV-1. Thus, we further distinguish between resting and activated virus specific T helper cells. The model is given by the following set of ordinary differential equations.

$$\dot{s} = \xi - fs - rsy, \tag{11.13}$$

$$\dot{x}_1 = rsy - d_1 x_1 - \beta_1 x_1 y, \tag{11.14}$$

$$\dot{x}_2 = \lambda - d_2 x_2 - \beta_2 x_2 y, \tag{11.15}$$

$$\dot{y} = y(\beta_1 x_1 + \beta_2 x_2) - ay - pyz, \tag{11.16}$$

$$\dot{z} = \frac{c x_1 y z}{\epsilon x_1 + 1} - bz. \tag{11.17}$$

where s denotes the number of virus specific resting T helper cells, x_1 the population of activated virus specific helper cells, x_2 the general population of T helper cells (that are not specific to the virus in question), y the number of infected cells, and z the CTL response. Resting specific T helper cells are are produced with a rate ξ, die with a rate fs, and become activated by virus with a rate rsy. Activated specific CD4 T cells die with a rate $d_1 x_1$ and become infected with a rate $\beta_1 x_1 y$. The general population of helper cells are produced with a rate λ, die with a rate $d_2 x_2$ and become infected with a rate $\beta_2 x_2 y$. Infected cells die with a rate ay and are killed by the CTL with a rate pyz. The CTL response grows with a rate $c x_1 y z/(\epsilon x_1 + 1)$, and the response decays with a rate bz. Since the focus of this model is not on the CTL, but on the helper cell responses, the CTL response is described by a simplified equation. CTL expansion is a saturating function of the level of specific helper cells, as described in Chapter 4, and the distinction between memory CTLp and effector CTL has been omitted.

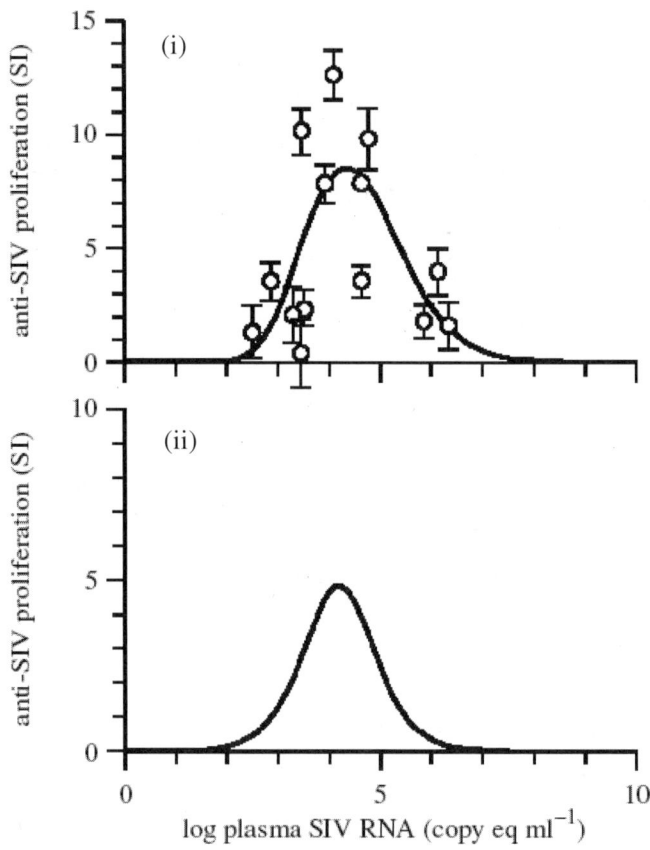

Fig. 11.6. Observed (i) and predicted (ii) relationship between viral load and SIV specific CD4 T cell proliferative responses. These are a measure of the amount of T cell help available. Data taken from Wodarz et al. Predictions are derived from model (11.13–11.17). We observe a one-humped distribution.

A complete analysis of this model will not be presented here and the interested reader is referred to [Wodarz et al. (2000a)]. We use this model to examine how the level of specific CD4 T cell responses correlates with virus load. As shown in Fig 11.6, we observe a one-humped relationship. That is at high virus load, the helper cell response is low. This is because most helper cells are infected and killed by the virus. That is, the virus has the upper hand. As the virus load is reduced, the level of the helper cell response increases. This is because a lower virus load corresponds with less infection and killing of the helper cells, and thus with less immune impairment. This negative correlation between virus load and the level of helper cell responses continues until a peak is reached. Now we enter a parameter region where lower virus loads correlate with reduced levels of helper cell responses. In other words, there is a positive correlation between virus load and the level of helper cell responses. The reason is that now immune control dominates over immune impairment. As the CTL response becomes stronger, virus load is suppressed to very low levels, and this removes the amount of antigenic stimulation required to maintain high numbers of specific helper cells. The lower the virus load, the lower the degree of antigenic stimulation, and the lower the specific helper cell responses. Thus, there are two basic parameter regions: if virus-induced immune impairment is the driving force of the dynamics and virus load is relatively high, then there is a negative correlation between virus load and helper cell responses. On the other hand, if immune control is sufficiently strong and the level of immune impairment is low, then there is a positive correlation between virus load and the level of helper cell responses. A low level of specific helper cells can therefore indicate either an impaired response where the virus has the upper hand, or a strong response where the immune system has the upper hand.

These model predictions have been confirmed with data from SIV infected macaques (Fig 11.6). Like HIV, the simian form of the virus infects and kills CD4 T helper cells. The typical course of disease shows a relatively high setpoint virus load after the acute phase, followed by gradual progression to AIDS over time. As will be explored in detail in Chapter 12, regimes of early therapy can induce varying degrees of CTL-mediated control in SIV infected macaques. These experiments have produced monkeys with a range of setpoint virus loads, from very low to levels which are typical of a progressing animal. The SIV specific CD4 T cell responses were measured in these animals, and plotted against virus load (Fig 11.6). Consistent with model predictions, a one-humped correlation was observed. Among animals showing the highest levels of virus control, the correlation was positive. Among animals showing less efficient virus control and higher loads, the correlation turns negative. This confirms the theoretical notion that the strength of the helper cell response does not necessarily correlate with the number of helper cells observed. Low numbers of helper cells can indicate either very efficient, or very inefficient control.

11.7 Summary

This chapter has explored the dynamics of virus infections that actively impair their specific helper cell responses. The most basic model indicated the existence of two alternative outcomes: CTL-mediated control versus CTL extinction. According to the model, a high replication rate of the virus promotes CTL extinction, and for intermediate viral replication rates, the outcome of infection can depend on the initial conditions. A high virus load and a low initial number of T cells promotes CTL extinction. These predictions are supported by data from LCMV infected mice. A more elaborate model distinguished between helper-dependent and helper-independent CTL responses. In this model, there is no CTL extinction. Instead we observe a qualitatively similar outcome: lack of long-term virus control in the presence of an inefficient CTL response. The conditions that promote lack of control are the same as those that promote CTL extinction in the simpler model. A more generalized model suggests that the conclusions described here are robust and do not depend on the exact mathematical formulation of immune impairment process. Finally, we pointed out that the level of the immune response does not necessarily correlate with the strength of the response. Low levels of specific T helper cell responses can be observed either if there is significant immune impairment and loss of virus control, or if virus control is very efficient.

12

Boosting Immunity against Immunosuppressive Infections

Chapter 11 explored the topic of virus-induced suppression of specific immune responses. In particular, we concentrated on the importance of helper cell impairment because this is a central feature of several persistent human viruses, such as HIV, HBV, and HCV. As explained in chapter 4, helper cell responses can be crucial for the resolution of viral infections by CTL. Therefore, if the impairment of helper cell responses can be reversed by treatment, long-term immunological control of the infection could be achieved without the need for further therapy. Long-term immune mediated control is a particularly desirable outcome with these persistent human infections because drug therapy comes with severe side effects and is difficult to tolerate. The subject of this chapter is to explore how therapy can be used to achieve long-term immunological control of immunosuppressive infections.

The problem is illustrated best in the context of HIV infection. antiviral drug therapy has conferred significant benefit to HIV infected patients. Administration of drug cocktails consisting of three or more different drugs can reduce and maintain virus load below detection limit in many patients. Nevertheless considerable problems remain such as viral resistance, side effects and lack of compliance during prolonged therapy [Condra et al. (1995); Frost and McLean (1994); Larder et al. (1989); Richman (1994)]. Furthermore it is unlikely that combination therapy alone can eradicate HIV from infected patients because of long-lived infected cells and sites within the body where drugs may not achieve effective levels [Chun et al. (1997); Finzi et al. (1997); Perelson et al. (1997); Wong et al. (1997)]. Currently, the only possibility to prevent the development of disease in patients is life-long therapy. It is not clear, however, for how long patients can tolerate these aggressive treatment regimes. Hence there is considerable interest in searching for therapy regimes that may reduce virus load and restimulate immune responses, thereby turning the balance between HIV and the immune system in favor of the immune system. Lessons can be learnt from so called long-term nonprogressors [Harrer et al. (1996a); Harrer et al. (1996b)]. These are HIV infected individuals who have not developed any sign of disease for as long as 15-20 years after infection. long-term nonprogressors are characterized by very low levels of virus load and high levels of HIV specific helper cell responses and CTL responses. In contrast, typically progressing patients are characterized by low levels of specific helper cell and CTL responses, already early in the course of infection [Rosenberg et al. (1997); Rosenberg et al. (1999)].

Although HCV is a very different virus compared to HIV, some of the immunological profiles appear similar [Lechner et al. (2000c)]. A fraction of patients clear HCV from the blood, while the rest of the patients develop a persistent infection that culminates in liver disease as long as 10 to 20 years after infection. Patients who clear the virus from the blood are characterized by substantial CD4 helper cell responses and sustained CTL responses against the virus. Patients who develop persistent infection are characterized by the absence of significant helper cell responses and the absence of sustained CTL responses [Day et al. (2002)]. Patients with persistent infection are treated with a combination of antiviral drugs and interferon [Barnes et al. (2002)]. In some patients, a course of therapy can result in sustained suppression of the virus after cessation of treatment, but in others, virus rebound is observed. Clinical data suggest that successful treatment correlates with the boosting of HCV specific immunity during therapy. Therefore, we are faced with a similar problem: given that HCV can impair specific T cell responses, how can we use treatment to reverse this impairment, and to induce long-term immunological control? HCV infection can be considered to be a less difficult case in this regard: In contrast to HIV, HCV does not kill T cells and the overall extent of immune impairment might be less.

In the following, we draw on the mathematical models from chapter 11. We review how they have been used to explore conceptual therapy regimes that can result in reversal of immune impairment and control of immunosuppressive infections. We show how the models can be applied to experiments with SIV infected macaques and to clinical data from HCV infected patients. We finish the chapter by discussing the concept of "structured treatment interruptions" that have been subject of much debate recently.

12.1 Basic Properties of Immune Impairment

Here we summarize essential results about immune impairment that are important in the context of therapy. This is based on the models presented in chapter 11. Again, we concentrate on the impairment of specific CD4 T helper cell responses. As reviewed in Chapters 3 and 4, CD4 T cell help is crucial for the generation of memory cell responses. Memory cell responses, in turn, may be crucial to ensure long-term virus control and resolution of infection (Chapter 3). HIV infection can result in the death of HIV specific CD4 T cells [Klenerman et al. (1996)]. HCV infection results in the impairment of virus specific helper cell responses, although the virus does not appear to kill the cells, and the mechanism underlying this impairment is unclear [Barnes

12.1 Basic Properties of Immune Impairment

et al. (2002)]. Therefore, both HIV and HCV infection might be characterized by the lack of efficient memory CTL responses that are maintained in the absence of antigen, and this could account for persistent infection and the eventual development of pathology. CTL responses that are observed in HIV infected patients may be suboptimal or "helper-independent" CTL, maintained by constant antigenic drive as a result of persistent replication. This is supported by clinical data [Kalams et al. (1999); Ogg et al. (1999)]: CTL responses are observed to decline to insignificant levels when virus is suppressed to low levels by antiviral drugs. This indicates that they are not long-lived in the absence of antigen and therefore do not share this memory phenotype. An aim of therapy should thus be the restoration of a sustained and efficient CTL memory response that can control the virus in the long-term.

In order to explore such therapy regimes, we will use the simplest and most general immune impairment model presented in chapter 11 (11.5–11.6) because this is most robust and independent of the exact form of the equations. It describes the interactions between an immunosuppressive infection y and the T cell response z:

$$\dot{y} = ygr(y) - yz,$$
$$\dot{z} = zf(y).$$

Although this model has been explained in Chapter 11, we will recapitulate the most important properties here because it is of central importance to the arguments presented in the current chapter. As explained in chapter 11, $gr(y)$ is a general function that describes the growth rate of the infection, and $f(y)$ is a general function that describes the expansion of the T cell response. For mathematical details, see Chapter 11. The basic assumption of the model is that intermediate virus loads stimulate the T response while higher virus loads impair the T cell response. We are interested in the effect of the viral replication rate r on the outcome of infection. This is because this parameter is reduced by antiviral therapy. Increasing the viral replication rate from low to high, we observe the following outcomes.

- If the replication rate is very small, the infection cannot be maintained. This outcome is denoted by S_0.

- If the replication rate of the virus crosses a threshold, an infection can be established but the amount of antigenic stimulation is still too low to trigger a sustained T cell response. The outcome is low virus load is the absence of immunity. This is denoted by S_n.

- If the replication rate crosses another threshold, levels of antigenic stimulation are sufficient to trigger a sustained T cell response. We observe T cell mediated control of the infection. This is denoted by S_i

- As the replication rate of the virus crosses a final threshold, levels of antigen can become relatively high such that immune impairment can outweigh the degree of antigenic stimulation. In this situation two alternative outcomes are possible: low virus load controlled by a sustained T cell response; and high virus load in the absence of a sustained T cell response. Which outcome is attained depends on the initial conditions. If the host is naive and has a low initial number of T cells, the likely outcome is loss of control. This is also promoted by a high initial virus load. On the other hand, if the initial number of T cell is already elevated, then the likely outcome is maintenance of the T cell response and control of the infection. This is also promoted by a low initial virus load.

In the following sections we explore CTL dynamics during antiviral therapy and investigate how therapy can be used to restore sustained CTL-mediated control of the infection.

12.2 T Cell Dynamics during Therapy

In this section we explore the dynamics of T cell responses during antiviral therapy. We assume that the system has approached the immune impairment outcome where the virus population is not controlled in the long-term. Because in this parameter region, both the immune impairment and the immune control outcome are stable, a temporary phase drug treatment can shift the dynamics into a domain of attraction where the T cell response retains control of the virus after the cessation of therapy. The bistability properties of the model are illustrated in Fig 12.1. Below line L, the system moves to the impairment outcome. Above line L, the system moves to the control outcome. Line L, which separates the two outcomes, is determined mainly by the number of T cells. Therefore, starting from the impairment outcome and a low number of T cells, the aim of therapy should be to push the number of T cells above line L. Once this has been achieved, therapy can be stopped and virus control will be maintained.

We assume that the infection is in the bistable parameter region (iv), and that the virus equilibrium has been attained. We ask how a single phase of therapy can establish sustained immune-mediated control of the infection. During therapy, the replication rate of the virus r is reduced in the model. The amount of reduction corresponds to the efficacy of the drugs. Upon cessation of therapy, the parameter r is reset to its pretreatment value. The CTL dynamics during drug therapy are as follows.

During therapy, the system converges toward a new equilibrium that is determined by the efficacy of the drugs. As explained above, sustained viral suppression will be only achieved upon cessation of treatment if therapy has moved the level of the immune response in the domain of attraction of the

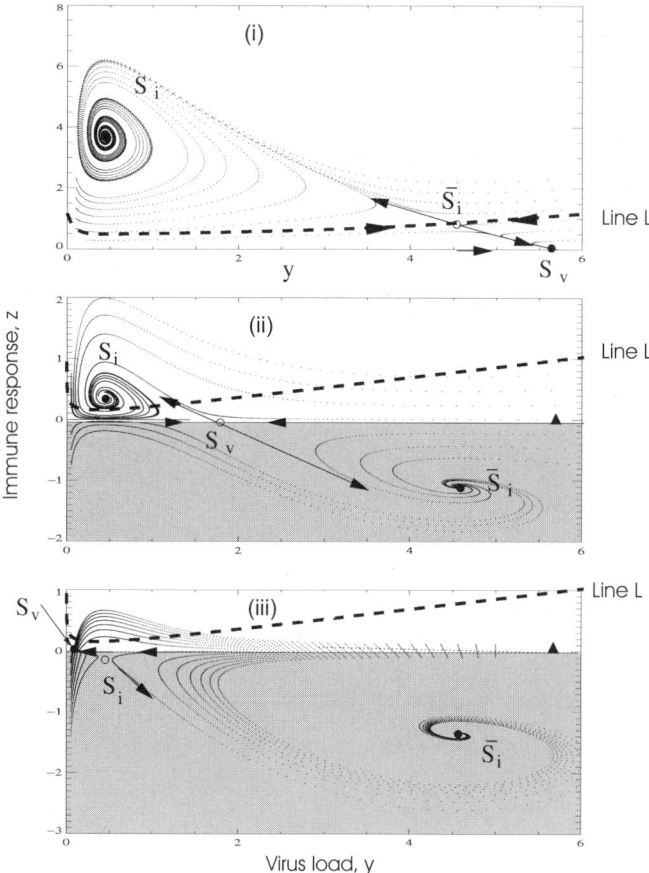

Fig. 12.1. Phase space diagram showing the behavior of the model. Equilibria are indicated, and the arrows show the equilibrium outcomes toward which the system moves over time depending on the starting conditions. Stable equilibria are indicated by closed circles. Unstable equilibria are indicated by open circles. Shading indicates the biologically irrelevant region where the immune response assumes negative values. The dashed lines indicate line L, the pretreatment separatrix. (i) The pretreatment regime and behavior of the model in the bistable parameter region. (ii), (iii) Therapy-mediated reduction of the viral replication rate r by an intermediate and a stronger amount, respectively. The flow lines are generated for a particular model that is an example of the class of models examined in our general framework (11.5–11.6)(see [Komarova et al. (2003)] for details and parameters).

immune control equilibrium (above line L). In this context, it is important to point out that there is a trade-off between the amount of immune impairment and antigenic stimulation in the expansion of immune responses. Treatment has to be efficient enough to sufficiently reduce the degree of immune impair-

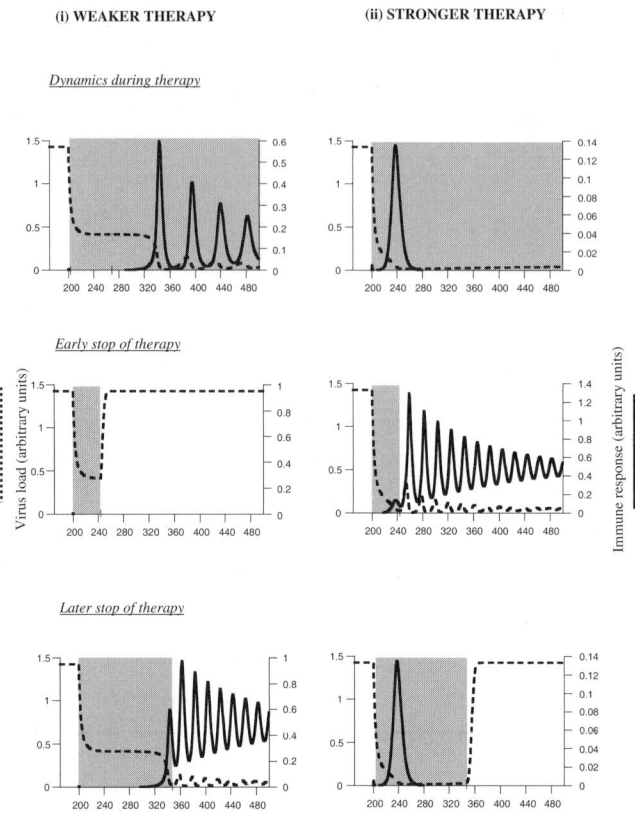

Fig. 12.2. Simulation of therapy assuming relatively weak and relatively strong therapy. Phases of treatment are indicated by shading. **(i)** With weaker therapy, therapy should be stopped after a time threshold, once the level of immune responses has significantly risen. **(ii)** With strong therapy, therapy has to be stopped early, before the temporary immune expansion has vanished. The graphs are generated for a particular model that is an example of the class of models examined in our general framework (11.5–11.6) (see [Komarova et al. (2003)] for details and parameters). Parameters were chosen as follows: $r = 3.5$, $k = 10$, $a = 3$, $p = 1$, $c = 12$, $\epsilon = 5$, $u = 3$, $b = 0.3$. For (i) $r = 3.13$ during therapy; for (ii) $r = 3.013$ during therapy.

ment, allowing the immune response to expand. However, if treatment is too efficient, not only immune impairment, but also the degree of antigenic stimulation is diminished. This reduces the amount of immune expansion. The following cases of increasing drug efficacy are discussed (the dynamical behavior of the model is summarized in Fig 12.1, and the corresponding time series are shown in Fig 12.2).

For treatment to have any effect, the drugs must have a minimum level of efficacy so that the viral replication rate is reduced from region (iv) into region (iii). In this case, the only stable equilibrium during treatment is the establishment of an immune response and virus control, S_i. (Fig 12.1ii). The system will move from the pretreatment equilibrium toward the treatment equilibrium. During this process the immune response expands (Fig 12.2i). After a time $t > t_{\min}$, the level of the response moves above line L. Hence, in principle, after the duration of therapy has crossed this time threshold, treatment can be stopped and sustained viral suppression will be achieved (Fig 12.2i). Note that the time it takes the immune response to cross line L can be long, because the system has to pass an unstable saddle equilibrium S_v (Fig 12.1ii). This time period can be shortened if in addition to drug treatment, the number of immune cells are augmented by immunotherapeutic approaches. Note in Figs 12.1ii and 12.2i that the approach to the treatment equilibrium is oscillatory. Conditions for when such oscillations are expected to occur have been defined in [Komarova et al. (2003)]. If they do occur, they may have different implications depending on the efficacy of the drugs. We have two possible scenarios. If therapy is less efficient, the treatment equilibrium (S_i) lies high above the line L, and the oscillations do not pose a problem. If therapy is more efficient, the treatment equilibrium (S_i) lies above but closer to line L. After the immune response has risen to a peak, the oscillations can take the response temporarily below line L before the treatment equilibrium is reached (Fig 12.1ii). Therapy should not be stopped in the phases when the immune response goes through troughs.

If therapy is even more efficient, the treatment equilibrium lies below line L (greatly reduced virus load cannot maintain immunity during treatment). This occurs if the replication rate of the virus is pushed from parameter region (iv) to region (ii). As shown in Fig 12.1iii, the only stable outcome during treatment is the presence of the virus (at reduced levels) in the absence of sustained long-term immunity, S_v. Since the treatment equilibrium lies below line L, cessation of therapy when equilibrium has been reached will result in rebound of the virus (Fig 12.2ii). However, on the way to this treatment equilibrium, the number of immune cells can temporarily rise above line L soon after start of therapy. Subsequently, it declines to low levels (Fig 12.2ii). This is because during the initial phase of treatment, immune impairment has been reduced, but levels of virus load are still high enough to stimulate immune expansion. Once virus load is reduced further by the drugs, this initial immune expansion diminishes due to lack of antigenic stimulation. Thus, to achieve sustained viral suppression in this case, treatment must not be continued for too long: treatment has to be stopped early while the level of immune cells is still high enough and above line L (Fig 12.2ii). Note, however, that the peak of the response during this temporary phase of expansion can lie below line L (Fig 12.1). In this case virus rebound will always be observed when therapy is stopped. Such an outcome is promoted by very strong drug-mediated suppression of viremia, and/or the absence of a sufficient number of reactive immune cells.

To summarize, the analysis has given rise to the insight that a single phase of drug treatment during chronic infection can result in long-term control of the virus. We have described the following relationship between efficacy, duration, and success of therapy (summarized in Fig 12.2). Therapy has to be efficient enough to reduce the rate of viral replication r at least from parameter region (iv) to region (iii). Within this constraint, significant immune responses develop during therapy only if suppression of viremia is relatively weak because this ensures the presence of sufficient antigenic drive. Control is maintained if therapy is stopped after a defined time threshold, once immunity has peaked and become established. On the other hand, if treatment is stronger, immune responses peak early after initiation of treatment and subsequently decline because the level of antigenic drive is diminished. Long-term control now requires an early stop of treatment, before immunity has significantly declined. Treating too long will result in failure. A single phase of treatment will not lead to sustained immunological control if drug-mediated suppression of viremia is too strong, or if the number of reactive immune cells is too low.

12.3 Application: Early Treatment of SIV/HIV

Experimental data from SIV infected macaques that received drug therapy during the early stages of primary infection [Lifson et al. (2000); Lifson et al. (2001); Wodarz et al. (2000a)] (Fig 12.3) are consistent with the model results. Fig 12.4 shows model simulations that should be compared to the experimental data on SIV infected macaques in Fig 12.3 and described as follows. As shown in Fig 12.3i, in the absence of any treatment, viral load increased to a peak value at approximately two weeks post inoculation, then declined, eventually equilibrating at levels typical of progressing SIV infection ($< 10^6$ SIV RNA copy Eq ml^{-1} of plasma). SIV specific lymphoproliferative responses, which indicate CD4 T helper cell function, were notably limited in this animal. In striking contrast, Fig 12.3ii shows results for an animal in which antiretroviral treatment was initiated 24 hrs post inoculation and continued for 28 days. In this animal, no viral RNA was detected in the plasma during the treatment period, or during a six week followup period after discontinuation of treatment. However, SIV specific lymphoproliferative responses were demonstrated during this treatment period, indicative of immunological sensitization of CD4 T cells responding to virus. Continuing cellular immunological sensitization, in the absence of measurable plasma viremia or seroconversion, probably reflects responses to replicating virus that was present at very low levels, or in an anatomically contained site. When the animal showed no evidence of viral replication during the first six weeks after discontinuation of treatment, it was rechallenged with a second infectious inoculum of the same pathogenic SIV isolate. Strikingly, no plasma virus was detected, although SIV specific lymphoproliferative responses increased transiently during the rechallenge. These results suggest that the initial treatment regimen had not only prevented the establishment of persistent productive infection, but has also conferred resistance to subsequent direct, intravenous rechallenge with a known infectious inoculum of a highly pathogenic SIV isolate. Peripheral blood mononuclear cells (PBMC) from the animal were readily susceptible to SIV infection in vitro, demonstrating that absence of productive established infection *in vivo* was not due to any intrinsic resistance to infection at the cellular level.

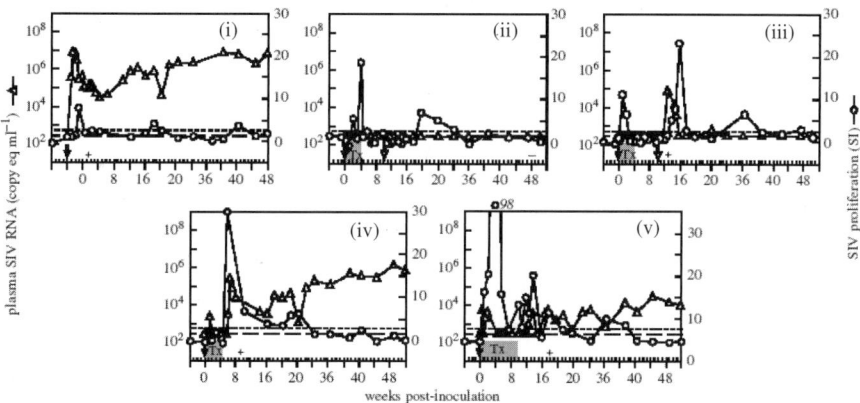

Fig. 12.3. Representative data illustrate different scenarios of viral and host dynamics in SIV infection. Data taken from [Wodarz et al. (2000a)] and [Lifson et al. (2000)]. (i) Plasma viremia (triangles) and anti-SIV proliferative responses (circles) for an untreated rhesus macaque infected intravenously. Dashed lines represent the threshold sensitivity of the virus load assay and the background of the anti-SIV lymphoproliferation assay. The solid arrow on the x-axis indicates the time of inoculation. The plus sign indicates the time of seroconversion to SIV antigens. (ii) Result of a rhesus macaque that received 28 days of treatment with tenofovir (PMPA), beginning 24 hrs post inoculation. Graphing conventions are as for (i), except that shaded box labeled Tx indicates the interval of tenofovir treatment, the minus symbol on the x axis indicates that the animal did not seroconvert, and the open arrow on the x axis indicates the time when the animal received a second intravenous challenge with the same SIV isolate. (iii) Results from another macaque that received 28 days of treatment beginning 24 hrs post infection. (iv) Results of a rhesus macaque that received 28 days of treatment and was started 72hrs post inoculation. (v) Results for a final animal that received 63 days of treatment beginning 72hrs post inoculation.

Fig 12.3iii shows a third animal in which virus was not detected in the plasma following the initial challenge, or in the immediate followup period, but was detected following the rechallenge. Strikingly, this post challenge peak was self limited, declining below the threshold of measurement, in conjunction with boosting of anti-SIV proliferative responses. After an additional, more blunted peak of plasma virus, which declined in conjunction with a boosting of proliferative responses, plasma viremia declined to below the threshold of measurement and has remained undetectable for months.

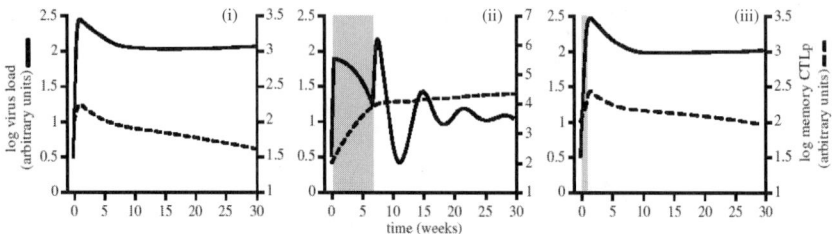

Fig. 12.4. Qualitatively different outcomes predicted by a mathematical model of anti-SIV immune responses. Simulations are based on a variant of the models discussed here, see [Wodarz et al. (2000a)] for details and parameter values. Shaded areas represent intervals of antiviral therapy. (i) In the absence of treatment, virus-induced impairment of CD4 helper cells precludes the development of sustained high levels of memory CTL, resulting in the lack of efficient virus control. Treatment during the primary phase of infection facilitates the establishment of a sustained CTL response by inhibiting excessive viral impairment of the helper CD4 helper cell responses, while still allowing antigenic sensitization. (ii) With a sufficiently long treatment period, begun shortly after inoculation, a sustained CTL response is established, facilitating sustained control of viral replication, with damped oscillations of viral suppression and reemergence. (iii) When brief treatment is begun shortly after inoculation, CTL respnses develop, but high level replication of virus emerging after treatment discontinuation results in the absence of a sustained CTL response.

Results from two other animals illustrate the potential impact of varying the delay before initiation of treatment, and the duration of treatment. As shown in Fig 12.3iv, when the same 28 day treatment regimen was not started until 72 hrs post inoculation, measurable levels of SIV RNA were detected transiently during the initial portion of the treatment period. While plasma viremia declined below detection levels during the treatment interval, it rapidly reemerged following drug discontinuation. SIV specific lymphoproliferative responses were not measured during the treatment period in this animal, but appeared rapidly in conjunction with the peak of reemergent viremia following treatment discontinuation. The peak level of viremia reached after stopping treatment was lower than that typically seen with untreated primary infection, and subsequently decreased to below 10^4 SIV RNA copy eq ml^{-1}. As viral load fell, so did SIV specific proliferative responses; with subsequent increases in viral load, SIV specific proliferative responses increased. However, with continued followup, SIV specific proliferative responses eventually no longer increased in association with increases in viral load, consistent with cumulative depletion or functional compromise of the responding cell pool. In conjunction with this, viral load rose in a manner that suggested loss of immunological control of viral replication, equilibrating at a level typical of progressive SIV infection.

Results from a final animal showed that extending the duration of post inoculation antiretroviral treatment appeared to compensate, at least in part, for its delayed initiation. As shown in Fig 12.3v, delaying the start of treatment until 72 hrs post inoculation was again associated with the presence of measurable plasma viremia during the initial portion of the treatment period. With treatment, plasma viremia declined to below measurable levels, and remained there for the duration of the eight week treatment period. However, strong SIV specific proliferative responses were demonstrated during the treatment period. Upon discontinuation of treatment there were several apparent cycles of blunted peaks of reemergent viremia that spontaneously declined, often below the levels of detection without any further intervention. Cyclical peaks and decreases in SIV specific proliferative responses were also observed, with the system eventually equilibrating at levels of plasma viremia typically associated with slowly progressive disease.

Further work has demonstrated that the protection achieved by early post inoculation therapy is broad and mediated by CTL. Macaques that controlled SIV for a year after the post inoculation treatment were rechallenged with a heterologous and more virulent strain of SIV [Lifson et al. (2001)]. While infection generally resulted in a self limited peak of viremia, the virus was reduced below the level of detection. Therefore, the immunological protection was not strain specific. In addition, virus load rose sharply when CTL were depleted from the animals with antibody treatment. This demonstrates that CTL contributed significantly to the suppression of viremia.

What are the implications of these results for HIV infected patients? A single phase of treatment has so far never resulted in improved immunological control in patients. When therapy is stopped, virus load usually returns to pretreatment levels. There are two important differences between the experiments described above and the situation with HIV infected patients. In the macaque experiments, treatment was started very early, within hours or a few days post inoculation. The experiments have shown that even if the start of treatment is delayed to 72 hours, compromised virus control is observed. HIV infected patients are at best diagnosed weeks after infection. By that time the relevant immunological specificities might already be killed off by the virus. Another important difference is the drug regimen used. The macaques were treated with a single drug (PMPA), which is in fact less efficient at suppressing viral replication than the human drugs. The models suggest that less effective suppression of virus replication might promote immunity better because it provides sufficient low level antigenic stimulation to induce the response.

With HIV infected patients, a combination of three (or more) efficient drugs are used. This might lead to strong and quick suppression of the virus and very little antigenic stimulation for immunity to develop. In support of this notion, clinical data have shown that patients with intermittent blips of viremia during treatment show higher levels of CTL during treatment than patients that suppress viremia more efficiently [Ortiz et al. (2002)]. Thus, the direct application of these results to HIV infected patients is limited. The value of the results, however, is to show that there are two alternative outcomes of the infection - progression and long-term control - and that treatment can in principle switch the outcome from lack of long-term control into a state of long-term nonprogression. The results are also valuable in a wider context involving therapeutic vaccination approaches.

12.4 Application: Treatment of HCV Infection

We discuss our theory also in the context of HCV infection. Impairment of specific helper cell responses has been clearly documented, and the virus is not thought to destroy the T cells [Barnes et al. (2002); Lechner et al. (2000b); Lechner et al. (2000c)]. Therefore, the relevant immune cells that need to be stimulated may not be absent during chronic infection. Two types of dynamics can be observed in early HCV infection. A small fraction of patients clear the virus from the blood. This outcome is associated with the development of strong CD4 helper cell responses and sustained CTL responses that persist in the long-term following resolution of infection. The rest of the patients develop chronic productive infections. This outcome is associated with weak CD4 helper cell responses and lack of sustained CTL responses.

Data from treated HCV infected patients [Barnes et al. (2002)] support the model predictions about the immune response dynamics during drug treatment. This study looked at the dynamics of virus and the specific CD4 T cell responses during a single phase of treatment in 15 subjects. These subjects had persistent viremia and lack of significant CD4 T cell responses to the virus before therapy. Upon cessation of treatment, a fraction of these subjects were characterized by virus rebound, while the rest showed long-term control of the infection below detectable levels.

The T cell dynamics during treatment support the model simulations [Barnes et al. (2002)]: T cell responses first increase towards a peak and then decline to low levels. The exact timing of the rise and peak of T cell responses differs between patients with different outcomes of treatment. In patients with virus rebound, generally CD4 T cell responses temporarily increased and peaked after the start of treatment, followed by a decline to insignificant levels. In those patients, treatment was generally stopped after the T cell responses had peaked. In the light of our theoretical framework, an earlier cessation of treatment, while immune responses were around their peaks, might have resulted in containment of the infection. Patients characterized by long-term control after treatment showed, overall, later peaks of CD4 T cell responses, and therapy was stopped while immune responses were closer to these peak levels. After cessation of therapy the CD4 T cell responses were, in some cases, boosted to even higher levels, as suggested by our model simulations (Fig 12.2). These data confirm that a single phase of treatment can indeed result in long-term immunological control of an immunosuppressive infection, and that the observed dynamics are at least consistent with our theoretical framework. There is indication that similar dynamics can occur in HBV infection [Boni et al. (2001)].

12.5 Treatment Interruptions

An important result from the above discussion is that long-term immune control can be achieved by a single phase of therapy if the combination of timing and efficacy of treatment is optimized. This is in contrast to previous notions that suggested that special regimes, such as structured therapy interruptions, have to be used to boost immunity [Garcia et al. (2001); Lisziewicz and Lori (2002); Lori et al. (2000a); Lori et al. (2000b); Miller (2001); Montaner (2001); Ortiz et al. (1999); Ortiz et al. (2001); Rosenberg et al. (2000)]. Structured treatment interruptions are defined as multiple phases of treatment, separated by periods during which no treatment is received. According to our theoretical framework, therapy interruptions can be only helpful in a restricted parameter region. Namely, interruptions can be beneficial if the following two conditions hold: (1) drug-mediated viral suppression is too strong to allow sufficient antigenic stimulation, and (2) the number of reactive immune cells is reduced to very low numbers (however, the number of immune cells must be above a threshold for any therapy to work). In the limited circumstances when interruptions help, the dynamics are as follows.

Treatment interruption dynamics are illustrated in Fig 12.5. A single phase of therapy moves the system toward the treatment equilibrium; the immune response increases, but is not sufficiently boosted to cross line L. Upon cessation of therapy, the system moves back to the pretreatment equilibrium. However, as shown in Fig 12.5, the trajectory toward the treatment equilibrium upon start of therapy is different from the trajectory away from the treatment equilibrium when therapy is stopped. Namely, when treatment is stopped, the immune response expands further before it declines again. The reason is as follows. Upon cessation of treatment, the virus population grows. During this growth phase, virus load first attains levels at which the amount of antigenic stimulation outweighs immune impairment, and this results in immune expansion. Of course, as virus load grows further, immune impairment outweighs antigenic stimulation, resulting in an eventual decline of the response.

Based on this, therapy interruptions can work as follows. The first phase of treatment should be stopped while the immune response is around its maxi-

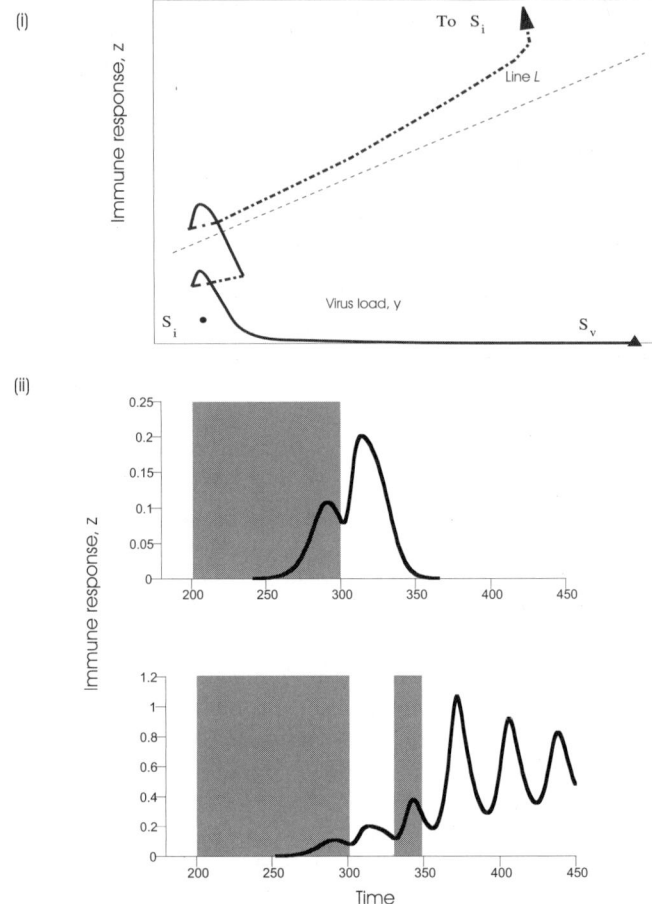

Fig. 12.5. Dynamics of structured therapy interruptions. (i) Phase space diagram. Solid line indicates treatment. Dashed line indicates phases off treatment. We start from the virus equilibrium, S_v, indicated by the black triangle. Interruptions can push the dynamics above line L. (ii) Time series presentation of these dynamics. While a single treatment phase fails to establish control, interruption therapy induces control. The plots are generated for a particular model that is an example of the class of models examined in our general framework (11.5–11.6) (see [Komarova et al. (2003)] for details and parameters). Parameter values were chosen as follows: $r = 0.7$, $k = 10$, $a = 0.1$, $p = 1$, $c = 0.6$, $\epsilon = 1$, $u = 0.075$, $b = 0.1$. During therapy, $r = 0.101$.

mum value (peak). During the off phase, the immune response will temporarily expand, as described above. When the immune response attains the maximum level during this off phase, therapy should be reapplied. This will result in further immune expansion that can push the response above line L. The second phase of treatment should, however, not be continued for too long: after the response peaks above line L, it will decline and fall back below line L. It is crucial that therapy is stopped while the immune response is still sufficiently high and above line L (Fig 12.5). If timing and duration of treatment are suboptimal, the immune response might not be pushed above line L by a single interruption. In this case, repeated therapy interruptions can increase the chances of success if the above scheme is continued.

Therapy interruptions received attention in the literature following some encouraging studies [Garcia et al. (2001); Lisziewicz and Lori (2002); Lori et al. (2000a); Lori et al. (2000b); Miller (2001); Montaner (2001); Ortiz et al. (1999); Ortiz et al. (2001); Rosenberg et al. (2000)]. An example is the treatment of a cohort of patients that were diagnosed during acute infection and presented symptoms. The patients received a number of treatment interruptions, and after about three interruptions, virus load remained at relatively low levels [Rosenberg et al. (2000)]. This improved control correlated with increased levels of HIV specific CTL. Other, larger scale and longer term studies, however, failed to show such positive effects of treatment interruptions on the level of immunological control of HIV [Kaufmann et al. (2004); Oxenius et al. (2002)]. One reason might be that these studies were performed in patients that were further advanced in the disease process, and not during acute infection. It is possible that the relevant immune cells that need to be stimulated had already been killed by that time, rendering a boost of immune responses a very difficult task.

12.6 Summary

This chapter has reviewed how antiviral drug therapy can be used to boost CTL memory in immunosuppressive infections that impair CD4 T cell help. We applied theory to HIV and HCV infections. The models, together with experimental and clinical data, show that a single phase of treatment can result in the restoration of CTL memory if the relationship between timing and the efficacy of drug treatment is optimized. In the context of SIV infection, we showed that the dynamics predicted by the equations are seen in experimental data from infected macaques that were treated shortly after inoculation. In the SIV/HIV system such early therapy is probably required because the virus quickly kills the necessary immune cells that need to react. Since HCV does not kill the helper cells, it is a more benign scenario in this respect. Clinical data suggest that a single phase of treatment in chronic infection can restore long-term virus control depending on the exact timing when treatment is stopped. We discussed the topic of structured treatment interruptions in the light of mathematical modeling and interpreted published clinical data.

13

Evolutionary Aspects of Immunity

This book has specialized in the dynamics of killer cell responses to viral infections. That is, we investigated the principles according to which CTL responses develop against viruses *in vivo*, how they fight the infection, and how viruses can fight back and subvert the CTL response. In other words, we concentrated on the interactions between populations of viruses and species of immune cells within the body of a host.

However, this discussion of CTL responses would be incomplete without evolutionary considerations. According to Theodosius Dobshanski, "nothing in biology makes sense except in the light of evolutionary biology". This chapter examines the interactions between CTL responses and infectious agents on an evolutionary level. The evolution of the immune system is a large subject and includes a wide variety of topics. In this chapter, we would like to concentrate on one particular aspect. We would like to discuss how the *in vivo* dynamics of CTL responses, and immune responses in general, influence the coevolution between pathogens and their hosts' immune system. This will be done in the context of immunological memory, one of the most important hallmarks of specific immune responses (Chapter 3).

Immunological memory can be defined as the prolonged persistence of elevated numbers of specific immune cells after the resolution of an infection (Chapter 3). Memory can protect against reinfection with the same pathogen. The duration for which protection lasts is not well defined and can vary from pathogen to pathogen. In some cases, protection can last a relatively long time, even for the life of the organism. In other cases, the phase of protection is relatively short lived. The most basic question is whether prolonged memory and protection is advantageous for the host. At first sight, this might seem like a trivial question, although there has been a fair amount of debate in the literature [Davenport (2000); Kundig et al. (1996a); Kundig et al. (1996b); Zinkernagel (2000b); Zinkernagel (2002a)]. The argument has been put forward that if an organism survives the first (primary) infection, it will also survive if it becomes infected a second time with the same organism [Zinkernagel (2000b); Zinkernagel (2002a)]. Hence, memory does not confer an advantage to the organism. This argument, however, does not take into account the fitness cost that can be associated even with a self limiting and acute infection [Davenport (2000)]. This fitness cost is greatly reduced if the organism if protected by immunological memory. For example, while the organism experiences symptoms from the infection, it is is more prone to being eaten by predators, and is less efficient at utilizing its natural resources. In addition to these arguments, however, there is additional complexity associated with this issue. It turns out that the duration of immunological memory can influence the fitness of virus strains that are characterized by different levels of virulence. Therefore, the duration of immunological memory can shape evolutionary processes in the virus population. This, in turn, can influence the evolution of the hosts' immune system. These complicated interactions will be examined with the help of mathematical models in the following sections. We start with a model that takes into account the interactions between a population of hosts and a single population of pathogens. Then, this model will be extended to include a diversity of pathogen strains or species in order to examine evolutionary processes.

13.1 A Single Population of Pathogens

We start with the description of a relatively simple epidemiological model that describes the interactions between a population of hosts and a single population of pathogens [Anderson and May (1991)] (Fig 13.1i). In contrast to all previous chapters, we do not consider the spread of the pathogen within a host. Instead, we consider the spread of the virus from host to host. This is because we are interested in the evolution of pathogens and their hosts. The model takes into account the following basic variables. Susceptible and uninfected hosts S, infected hosts I, recovered hosts that are protected against reinfection R, and a population of pathogens P. Uninfected and susceptible hosts are assumed to reproduce at a rate r and die at a rate d. Note that reproduction saturates at high host densities, and this is captured by the parameter ϵ. Hosts become infected by the pathogen at a rate β. Infected hosts are characterized by an elevated death rate a, reflecting pathogen-induced mortality. In addition, they are assumed to recover from infection at a rate α. Recovered hosts die at the same rate as uninfected individuals d, and they cannot be reinfected by the pathogen. This protection is not infinite, but is lost at a rate g. The model assumes that all host populations reproduce. The pathogen is, however, not transmitted vertically to the offspring. Moreover, it is assumed that offspring from recovered and immune hosts are once again susceptible to infection. (while antibody memory can be transferred from mother to child, this protection only lasts for a few months; T cell memory is not transferred from mother to offspring). Finally, pathogens can be released from the hosts into the environment at a rate k, and may decay in the environment at a rate u. The dynamics are formulated in terms of the following ordinary differential

13.1 A Single Population of Pathogens

equations that describe the development of these populations over time.

$$\dot{S} = \frac{r(S+I+R)}{\epsilon(S+I+R)+1} - dS - \beta SP + gR, \tag{13.1}$$

$$\dot{I} = \beta SP - aI - \alpha I, \tag{13.2}$$

$$\dot{R} = \alpha I - dR - gR, \tag{13.3}$$

$$\dot{P} = kI - uP. \tag{13.4}$$

(i) Basic dynamics

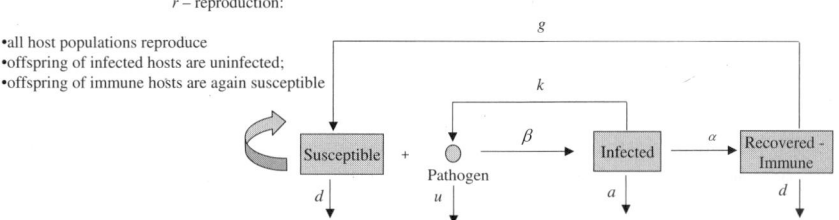

(ii) One host, two pathogens

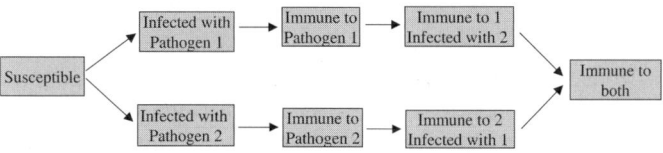

Fig. 13.1. Schematic representation of the mathematical models discussed in this chapter. (i) Basic epidemiological model that describes the spread of a pathogen through a host population (13.1–13.4). The model considers a single population of hosts and a single population of pathogens. (ii) Extended model that describes the interactions between a population of hosts and two populations (species or strains) of pathogens (13.5–13.14). For details, see text.

The model is characterized by two outcomes.

(i) The pathogen is not maintained and the host population remains uninfected ($S > 0, I = R = P = 0$).

(ii) The pathogen is maintained in the host population, and the susceptible, infected, and recovered hosts, as well as the pathogen population, settle at an equilibrium level. The expressions for this equilibrium are not written out here since they are long and complicated.

The pathogen is maintained in the population and outcome (ii) is observed if $\beta(r-d)/d\epsilon - a - \alpha > 0$.

We would like to understand in which direction the duration of memory and protection is expected to evolve. In the model, the rate at which immune hosts revert to being susceptible is given by the variable g. Hence, the average duration of memory is given by $G = 1/g$. In order to study in which direction the duration of memory (G) is expected to evolve, we analyze the competition between two populations of hosts that differ in the duration of memory. We examine whether a host population with a shorter or longer duration of memory can invade, and this exercise is repeated. This will tell us towards which duration immunological memory is expected to evolve. It is found that the host population with the longer duration of memory will always invade and outcompete the host population with the shorter duration of memory. Hence, the duration of memory is predicted to evolve to infinity $(G = \infty)$. Of course, in reality, the duration of memory would be finite because cells cannot live forever, and because of reproductive tradeoffs that are not included in the model. Therefore, we refer to this as the maximum memory outcome. Thus, in this simple setting, a host with a longer duration of immunological protection has a selective advantage and will survive.

13.2 Two Competing Pathogen Populations

Now, we consider a more complicated scenario. Instead of one pathogen population, we now consider two pathogen populations that infect the same host population (Fig 13.1ii). These could be two species of pathogens, or two strains of a given pathogen species. An important assumption is that the two pathogen species or strains are not immunologically cross-reactive. Therefore, hosts recovered from one pathogen are still susceptible to infection by the other. The model includes the following variables. Two populations of pathogens P_1 and P_2. The population of hosts infected with pathogen 1 are denoted by I_1, and hosts recovered and immune to pathogen 1 are denoted by R_1. Similarly, hosts infected by pathogen 2 are denoted by I_2, and hosts recovered and immune to pathogen 2 are denoted by R_2. Hosts immune to pathogen 1 are still susceptible to pathogen 2, and hosts immune to pathogen 2 are susceptible to pathogen 1. Thus, we have the following additional populations: Hosts recovered from pathogen 1 and infected with pathogen 2 I_{12}; hosts recovered from pathogen 2 and infected with pathogen 1 I_{21}. Hosts recovered and immune to both infections are denoted by R_{12}. For simplicity, it is assumed that hosts do not experience simultaneous multiple infections. The equations for the model are given by the following set of ordinary differential equations.

$$\dot{S} = \frac{rH}{\epsilon H + 1} - dS - \beta_1 SP_1 - \beta_2 SP_2 + g(R_1 + R_2 + R_{12}), \quad (13.5)$$

13.2 Two Competing Pathogen Populations

$$\dot{I}_1 = \beta_1 S P_1 - a_1 I_1 - \alpha_1 I_1, \tag{13.6}$$
$$\dot{R}_1 = \alpha_1 I_1 - dR_1 - gR_1 - \beta_2 R_1 P_2, \tag{13.7}$$
$$\dot{I}_2 = \beta_2 S P_2 - a_2 I_2 - \alpha_2 I_2, \tag{13.8}$$
$$\dot{R}_2 = \alpha_2 I_2 - dR_2 - gR_2 - \beta_1 R_2 P_1, \tag{13.9}$$
$$\dot{I}_{12} = \beta_2 R_1 P_2 - a_2 I_{12} - \alpha_2 I_{12}, \tag{13.10}$$
$$\dot{I}_{21} = \beta_1 R_2 P_1 - a_1 I_{21} - \alpha_1 I_{21}, \tag{13.11}$$
$$\dot{R}_{12} = \alpha_2 I_{12} + \alpha_1 I_{21} - dR_{12} - gR_{12}, \tag{13.12}$$
$$\dot{P}_1 = k_1 (I_1 + I_{21}) - u P_1, \tag{13.13}$$
$$\dot{P}_2 = k_2 (I_2 + I_{12}) - u P_2. \tag{13.14}$$

where the sum of the host population is given by $H = S + I_1 + R_1 + I_2 + R_2 + I_{12} + I_{21} + R_{12}$. The model is based on the simple one host one pathogen equations introduced in the previous section (13.1–13.4). Because there are now two pathogen strains, hosts that are immune to one pathogen can be infected by the other pathogen at a rate β. Note that this model includes the possibility for competition between the pathogens, because the two pathogen populations share the same host. The better a species of pathogen is at spreading through the host population, the more superior its competitive ability.

The model gives rise to the interesting finding that the duration of memory (G) can regulate the outcome of competition between the two pathogens (Fig 13.2). If the duration of memory is short and the population of immune hosts becomes susceptible again at a fast rate (low value of G), competition between pathogens is strong and the superior pathogen wins and excludes the inferior one. On the other hand, if the duration of protection is longer (higher value of G), competition between the two pathogens is weaker and coexistence of the pathogens can be observed (Fig 13.2). The longer the duration of protection, the higher the relative abundance of the competitively inferior pathogen (Fig 13.2). In other words, long lasting immunological memory allows a competitively inferior pathogen to persist and to be maintained in the host population. The reason is as follows. A long duration of memory results in the presence of hosts that are only susceptible to one, but not the other pathogen species. Therefore, the degree of interspecific competition is reduced relative to intraspecific competition. This results in coexistence. If, on the other hand, the duration of memory is relatively short, then the majority of hosts will be susceptible to both pathogen species. In this case, the degree of interspecific competition is much higher and competitive exclusion is observed. A relationship between immunity, cross-reactivity, and pathogen diversity has also been discussed in the context of influenza virus infection [Andreasen et al. (1997); Lin et al. (1999)].

Mathematically speaking, for the inferior pathogen to persist, the following condition has to be fulfilled: assuming that pathogen 1 is the superior competitor and pathogen 2 is the inferior competitor, the condition is

(i)

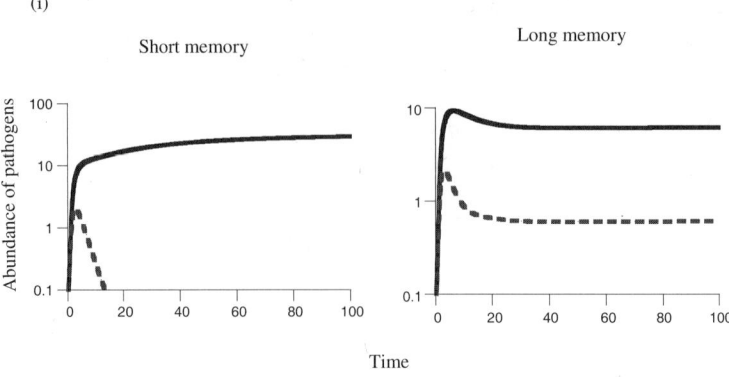

(ii)

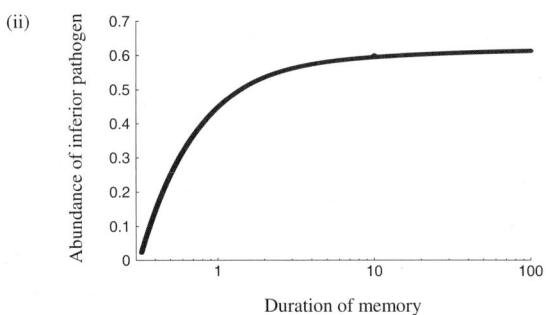

Fig. 13.2. Duration of memory and the competition between an inferior (dashed line) and a superior (solid line) pathogen. (i) If memory is short, the superior pathogen outcompetes and excludes the inferior one. If the duration of memory lies above a threshold, the inferior pathogen can coexist with the superior one. (ii) The longer the duration of memory, the higher the abundance of the inferior pathogen. Simulations are based on equations (13.5–13.14) Parameter values were chosen as follows: $r = 0.5$, $d = 0.01$, $\beta = 1$, $\alpha = 0.1$, $k = 1$, $u = 0.5$, $a_1 = 0.03$, $a_2 = 1$, (i) $g = 10$, (ii) $g = 0.01$.

$\beta_2 S^* + \beta_2 R_1^* > \alpha_2 + a_2$, where S^* and R_1^* are the equilibrium level of susceptible and recovered hosts assuming that the competitively superior pathogen P_1 is only present.

13.3 Pathogen Competition and the Evolution of Memory Duration

What are the implications of these findings for the evolution of immunological memory? The answer depends on the assumptions of the model. If the inferior

13.3 Pathogen Competition and the Evolution of Memory Duration

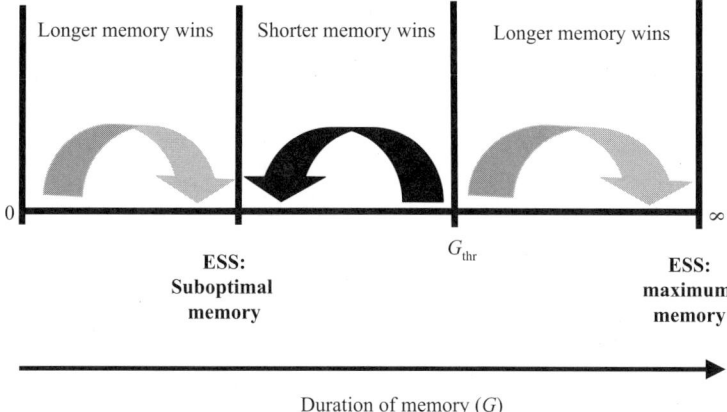

Fig. 13.3. Schematic representation depicting the evolution of immunological memory assuming the presence of two populations of pathogens that differ in their fitness (13.5–13.14). The pathogens differ in their virulence (rate of host killing), and we assume that we are in a parameter region where increased virulence correlates with reduced fitness. The system can evolve towards two different outcomes. The maximum memory outcome and the shorter or "suboptimal" memory outcome. Which outcome is achieved by evolution depends on the initial conditions; that is the initial duration of memory. If the simulation is started with an initial duration of memory that lies above G_{thr}, having a longer duration of memory is advantageous. Thus, evolution takes the system to the maximum memory outcome. If the simulation is started with an initial duration of memory that lies below G_{thr}, the system evolves towards the suboptimal memory outcome.

pathogen species is less virulent, persistence of this pathogen due to a longer duration of memory can be only advantageous for the host. The situation is, however, more complicated if the inferior pathogen is more virulent. In this case, the persistence of the inferior pathogen due to a longer duration of memory can be costly for the host population because the level of virulence is higher. From an evolutionary point of view this is the most interesting parameter region, and we will focus on this.

In this parameter region where the inferior pathogen is more virulent, the following arguments apply. While a long duration of memory is advantageous because it protects the host from reinfection, a short duration of memory can also be advantageous because it allows less virulent pathogens to exclude more virulent ones. We observe two outcomes towards which the system may evolve (Fig 13.3). One of the outcomes is maximum memory ($G = \infty$). The other outcome is a suboptimal and shorter duration of memory (smaller value of G). To which state the system evolves depends on the starting condition G_0 (Fig 13.3). If we start with a duration of memory that lies above a threshold ($G_0 > G_{\text{thr}}$), the system evolves towards maximal memory. If we start with

13 Evolutionary Aspects of Immunity

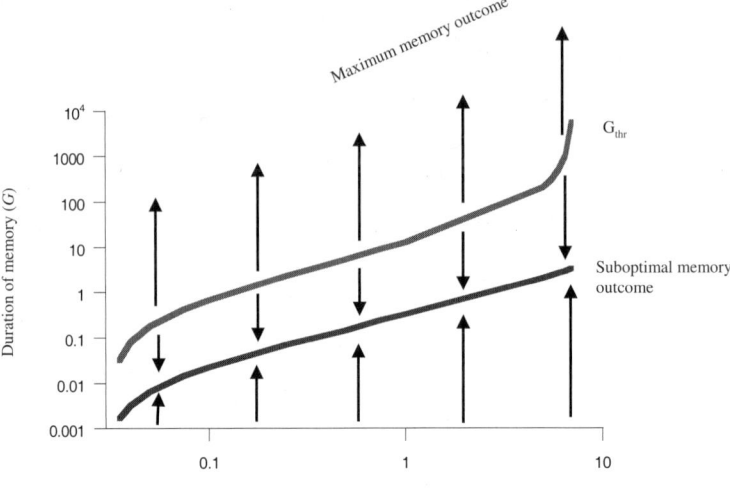

Fig. 13.4. Dependence of the value of G_{thr} - separating the two outcomes - on the degree of host killing by the more virulent pathogen. We assume that we are in a parameter region where increased virulence results in reduced fitness. Starting with a duration of memory that lies above G_{thr}, the system evolves to the maximum memory outcome (indicated by the arrows pointing upwards). Starting with an initial duration of memory that lies below G_{thr}, the system evolves to the suboptimal memory outcome, which is indicated by a line. If the degree of host killing by the more virulent pathogen is similar to that of the less virulent one, G_{thr} is low and the system is likely to evolve to the maximum memory outcome. If the degree of host killing by the more virulent pathogen is higher, G_{thr} is high and the system is likely to evolve to the suboptimal memory outcome. As the degree of host killing by the more virulent pathogen is increased, the duration of memory at the suboptimal outcome becomes longer, because less memory reduction is required to exclude more virulent pathogens. Plot is based on equations (13.5–13.14). Parameters were chosen as follows: $r = 0.5$, $d = 0.01$, $\beta = 1$, $\alpha = 0.03$, $k = 1$, $u = 0.5$, $a_1 = 0.03$, $a_2 = 1$.

a duration of memory that lies below the threshold ($G_0 < G_{\text{thr}}$), the system evolves to the state describing suboptimal memory. What is the initial duration of memory that separates the two outcomes (value of G_{thr})? As shown in Fig 13.4, it depends on the rate of host killing by the more virulent pathogen. At one extreme, the rate of host killing by the more virulent pathogen is similar to that of the less virulent pathogen. Fig 13.4 shows that the threshold duration (G_{thr}) separating the two outcomes is short. Therefore, the system is likely to evolve to maximum duration of memory. The reason is that the difference in virulence between the two pathogens is low. Thus, it does not pay to reduce memory in order to exclude the more virulent pathogen. At the opposite extreme, the rate of host killing by the more virulent pathogen is much higher than that of the less virulent one. Now, the threshold duration (G_{thr}) that separates the two outcomes is very high. In this case, the system is likely to evolve towards the suboptimal memory outcome. The reason is that the more virulent pathogen is characterized by a much higher rate of host killing compared to the less virulent one. It therefore confers a significant cost to the host population, and exclusion of the more virulent pathogen by means of memory reduction confers a significant advantage. Note, however, that the duration of protection at the suboptimal memory outcome becomes longer as the rate of host killing by the more virulent pathogen is increased. The reason is that a higher rate of host killing reduces the relative fitness of the more virulent pathogen, and thus less memory reduction is required to exclude it.

Note, however, that the two outcomes to which the system can evolve may not be stable states. Assume that evolution takes the system to the suboptimal memory outcome. This can result in the exclusion of a more virulent pathogen and reduction in pathogen diversity. As pathogen diversity is reduced, it will become advantageous again to evolve a longer duration of memory because this leads to lasting protection. As memory becomes longer, however, inferior pathogens may invade again. As a consequence more virulent pathogens can persist and pathogen diversity increases. In this scenario, it will once again pay to evolve towards a shorter duration of memory. Thus, we may expect the duration of memory to cycle over time.

13.4 Application to Immunological Observations

Following the theoretical discussion, we now try to relate the results to immunological data. This is a challenging task because much of the information required to link theory and data is currently not available. In order to test theory, the duration of protection against reinfection needs to be measured in the context of two types of pathogens. One pathogen needs to be genetically diverse and cocirculate as a collection of distinct serotypes; the other pathogen should be homogeneous. Based on theory, immunity against any strain/serotype of the diverse pathogen should be shorter than protection against the homogeneous pathogen. For this to be true, however, the following conditions need to hold: the serotypes should be characterized by differences in pathogenicity, and the more pathogenic strains should have reduced fitness relative to the less pathogenic strains. This is very difficult to quantify [Lipsitch and Moxon (1997)], and such information does not currently exist to our knowledge.

Nevertheless, it is interesting to consider rhinoviruses as an example [Groarke and Pevear (1999); Pevear et al. (1999)]. Rhinoviruses are the causative agents of the common cold. They cocirculate as a collection of many different serotypes. The primary infection results in the generation of IgA in nasal secretions and IgG in blood stream. Acquired immunity is type specific, and correlates with the level of mucosal IgA antibodies [Hendley (1999)]. (While other responses, such as CTL, are not necessarily type specific, their role in protection is uncertain). The titer of IgA antibodies, however, declines at a relatively fast rate, and protection is thought to only last for one or two years [Barclay et al. (1989); Gern et al. (1996)]. This is in contrast to the observation that many infections, as well as vaccines, can result in the presence of protecting antibodies for decades [Slifka and Ahmed (1998)]. It can be hypothesized that the shorter duration of memory is observed against rhinoviruses because it prevents the maintenance of pathogenic strains in the host population. Validation of this hypothesis would, however, require detailed investigations, as outlined above. This could have important implications for vaccination approaches: prolonging the duration of protection against rhinoviruses by vaccination might allow for the emergence of more pathogenic strains, resulting in a cost for the human population. It is an important question how the duration of memory can be modulated. For example, it has been observed that memory CTL (cytotoxic T lymphocytes) can persist for a very long time in the absence of antigenic stimulation [Murali-Krishna et al. (1999)], and this should apply to all infections. Interestingly, it has also been demonstrated that the number of memory CTL against a given pathogen can be reduced by an antigenically heterologous infection [Selin et al. (1999)] (see Chapter 6). Thus, if a host is exposed to many serologically distinct pathogens in a relatively short period of time, memory against the pathogens can be reduced and might last for a shorter period of time. It is interesting to consider the hypothesis that this could be adaptive for the host: it may allow a reduction in pathogen diversity and the extinction of more pathogenic strains. Other factors can also modulate the duration of immunological memory. The interactions between pathogens and the immune system can result in the impaired generation of memory and thus in shorter protection. If shorter protection is advantageous for the host, there will be no selection pressure to overcome this impairment. On the other hand, protection can be prolonged if memory is rebooted. This can occur if reinfection is not prevented and limited pathogen growth occurs. While symptoms may be absent, the limited pathogen growth could increase an otherwise declining memory cell population. These examples show that the duration of memory should not be considered as a fixed constant, but as a variable that can be modulated. The models presented here provide a framework in which to consider the topic of memory duration in more detail.

13.5 Summary

This chapter examined how immune response dynamics *in vivo* can be linked to the evolution of pathogens and their hosts. This was done by focusing on the duration for which hosts are protected against reinfection by the memory cells. If a population of hosts is faced with a single population of pathogens, a longer duration of protection is always advantageous for the host. The situation is different, however, if a given host population is infected by more than one species or strain of pathogen that are not immunologically cross-reactive. Now, the duration of memory can influence the competition between the pathogens. In specific circumstances, a relatively long duration of memory can allow the coexistence of a more virulent pathogen with a less virulent one. On the other hand, if the duration of memory is short, a more virulent pathogen can be driven extinct by the less virulent one. This gives rise to the following evolutionary patterns. If the duration of memory is short, pathogen diversity is expected to decrease, and a pathogen with a relatively low degree of virulence is likely to prevail. In this case, the host will evolve towards a longer duration of memory. This in turn allows an increase in pathogen diversity and the emergence of pathogen strains that are characterized by increased levels of virulence. This, however, is costly for the host, so the host's immune system is expected to evolve back towards shorter memory durations. Therefore, the evolutionary trajectories of the host and pathogens are expected to move in cycles.

References

Adorini, L., Appella, E., Doria, G.,Nagy, Z.A.: Mechanisms Influencing the Immunodominance of T-Cell Determinants. Journal of Experimental Medicine **168**, 2091-2104 (1988)

Ahmed, R., Butler, L.D.,Bhatti, L.: T4+ T helper cell function in vivo: differential requirement for induction of antiviral cytotoxic T-cell and antibody responses. J Virol **62**, 2102-6 (1988)

Ahmed, R.,Gray, D.: Immunological memory and protective immunity: understanding their relation. Science **272**, 54-60 (1996)

Ahmed, R.,Oldstone, M.B.: Organ-specific selection of viral variants during chronic infection. J Exp Med **167**, 1719-24 (1988)

Andersen, C., Jensen, T., Nansen, A., Marker, O.,Thomsen, A.R.: CD4(+) T cell-mediated protection against a lethal outcome of systemic infection with vesicular stomatitis virus requires CD40 ligand expression, but not IFN-gamma or IL-4. Int Immunol **11**, 2035-42. (1999)

Anderson, R.M.,May, R.M. Infectious diseases of humans. Oxofrd, England: Oxfors University Press(1991)

Andersson, E.C., Christensen, J.P., Marker, O.,Thomsen, A.R.: Changes in cell adhesion molecule expression on T cells associated with systemic virus infection. J Immunol **152**, 1237-45 (1994)

Andreasen, S.O., Christensen, J.E., Marker, O.,Thomsen, A.R.: Role of CD40 ligand and CD28 in induction and maintenance of antiviral CD8+ effector T cell responses. J Immunol **164**, 3689-97 (2000)

Andreasen, V., Lin, J.,Levin, S.A.: The dynamics of cocirculating influenza strains conferring partial cross-immunity. J Math Biol **35**, 825-42. (1997)

Arase, H.,Lanier, L.L.: Virus-driven evolution of natural killer cell receptors. Microbes Infect **4**, 1505-12 (2002)

Arnaout, R.A.,Nowak, M.A.: Competitive coexistence in antiviral immunity. J Theor Biol **204**, 431-41. (2000)

Bachmann, M.F., Kalinke, U., Althage, A., Freer, G., Burkhart, C., Roost, H., Aguet, M., Hengartner, H.,Zinkernagel, R.M.: The role of antibody concentration and avidity in antiviral protection. Science **276**, 2024-7 (1997)

Badovinac, V.P., Porter, B.B.,Harty, J.T.: Programmed contraction of CD8(+) T cells after infection. Nat Immunol **3**, 619-26. (2002)

Badovinac, V.P., Tvinnereim, A.R.,Harty, J.T.: Regulation of Antigen-Specific CD8(+) T Cell Homeostasis by Perforin and Interferon-gamma. Science **290**, 1354-1358 (2000)

Bangham, C.R.: Human T-cell leukaemia virus type I and neurological disease. Curr Opin Neurobiol **3**, 773-8 (1993)

Bangham, C.R.M., Hall, S.E., Jeffrey, K.J.M, Vine, A.M., Witkover, A., Nowak, M.A., Wodarz, D., Usuku, K., Osame, M.: Genetic control and dynamics of the cellular immune response to the human T-cell leukaemia virus, HTLV-1. Philosophical Transactions of the Royal Society, Series B **354**, 691-700 (1999)

Barclay, W.S., al-Nakib, W., Higgins, P.G.,Tyrrell, D.A.: The time course of the humoral immune response to rhinovirus infection. Epidemiol Infect **103**, 659-69. (1989)

Barnes, E., Harcourt, G., Brown, D., Lucas, M., Phillips, R., Dusheiko, G.,Klenerman, P.: The dynamics of T-lymphocyte responses during combination therapy for chronic hepatitis C virus infection. Hepatology **36**, 743-54. (2002)

Bartholdy, C., Christensen, J.P., Wodarz, D.,Thomsen, A.R.: Persistent virus infection despite chronic cytotoxic T-lymphocyte activation in gamma interferon-deficient mice infected with lymphocytic choriomeningitis virus. J Virol **74**, 10304-11 (2000)

Belz, G.T., Wodarz, D., Diaz, G., Nowak, M.A.,Doherty, P.C.: Compromised influenza virus-specific CD8(+)-T-cell memory in CD4(+)-T- cell-deficient mice. J Virol **76**, 12388-93. (2002)

Bergmann, C.C., Altman, J.D., Hinton, D.,Stohlman, S.A.: Inverted immunodominance and impaired cytolytic function of CD8+ T cells during viral persistence in the central nervous system. J Immunol **163**, 3379-87 (1999)

Betts, M.R., Ambrozak, D.R., Douek, D.C., Bonhoeffer, S., Brenchley, J.M., Casazza, J.P., Koup, R.A.,Picker, L.J.: Analysis of total human immunodeficiency virus (HIV)-specific CD4(+) and CD8(+) T-cell responses: relationship to viral load in untreated HIV infection. J Virol **75**, 11983-91 (2001)

Beverley, P.C.: Is T-cell memory maintained by crossreactive stimulation? Immunol Today **11**, 203-5 (1990)

Binder, D., van den Broek, M.F., Kagi, D., Bluethmann, H., Fehr, J., Hengartner, H.,Zinkernagel, R.M.: Aplastic anemia rescued by exhaustion of cytokine-secreting CD8+ T cells in persistent infection with lymphocytic choriomeningitis virus. J Exp Med **187**, 1903-20. (1998)

Bocharov, G., Ludewig, B., Bertoletti, A., Klenerman, P., Junt, T., Krebs, P., Luzyanina, T., Fraser, C.,Anderson, R.M.: Underwhelming the immune response: effect of slow virus growth on CD8+-T-lymphocyte responses. J Virol **78**, 2247-54 (2004)

Bonhoeffer, S., May, R.M., Shaw, G.M.,Nowak, M.A.: Virus dynamics and drug therapy. Proceedings of the National Academy of Sciences of the United States of America **94**, 6971-6976 (1997)

Bonhoeffer, S.,Nowak, M.A.: Pre-existence and emergence of drug resistance in HIV-1 infection. Proc R Soc Lond B Biol Sci **264**, 631-7 (1997)

Boni, C., Penna, A., Ogg, G.S., Bertoletti, A., Pilli, M., Cavallo, C., Cavalli, A., Urbani, S., Boehme, R., Panebianco, R., Fiaccadori, F.,Ferrari, C.: Lamivudine treatment can overcome cytotoxic T-cell hyporesponsiveness in chronic hepatitis B: new perspectives for immune therapy. Hepatology **33**, 963-71. (2001)

Borghans, J.A., Taams, L.S., Wauben, M.H.,de Boer, R.J.: Competition for antigenic sites during T cell proliferation: a mathematical interpretation of in vitro data. Proc Natl Acad Sci U S A **96**, 10782-7. (1999)

Borrow, P., Evans, C.F.,Oldstone, M.B.: Virus-induced immunosuppression: immune system-mediated destruction of virus-infected dendritic cells results in generalized immune suppression. J Virol **69**, 1059-70 (1995)

Borrow, P., Lewicki, H., Wei, X.P., Horwitz, M.S., Peffer, N., Meyers, H., Nelson, J.A., Gairin, J.E., Hahn, B.H., Oldstone, M.B.A., Shaw, G.M.: Antiviral pressure exerted by HIV-1-specific cytotoxic T lymphocytes (CTLs) during primary infection demonstrated by rapid selection of CTL escape virus. Nature Medicine **3**, 205-211 (1997)

Borrow, P., Tishon, A., Lee, S., Xu, J., Grewal, I.S., Oldstone, M.B., Flavell, R.A.: CD40L-deficient mice show deficits in antiviral immunity and have an impaired memory CD8+ CTL response. J Exp Med **183**, 2129-42 (1996)

Borrow, P., Tough, D.F., Eto, D., Tishon, A., Grewal, I.S., Sprent, J., Flavell, R.A.,Oldstone, M.B.: CD40 ligand-mediated interactions are involved in the generation of memory CD8(+) cytotoxic T lymphocytes (CTL) but are not required for the maintenance of CTL memory following virus infection. J Virol **72**, 7440-9 (1998)

Brehm, M.A., Pinto, A.K., Daniels, K.A., Schneck, J.P., Welsh, R.M., Selin, L.K.: T cell immunodominance and maintenance of memory regulated by unexpectedly cross-reactive pathogens. Nat Immunol **3**, 627-34 (2002)

Broussard, S.R., Staprans, S.I., White, R., Whitehead, E.M., Feinberg, M.B., Allan, J.S.: Simian immunodeficiency virus replicates to high levels in naturally infected African green monkeys without inducing immunologic or neurologic disease. J Virol **75**, 2262-75 (2001)

Bruno, L., Kirberg, J.,von Boehmer, H.: On the cellular basis of immunological T cell memory. Immunity **2**, 37-43 (1995)

Cabot, B., Martell, M., Esteban, J.I., Sauleda, S., Otero, T., Esteban, R., Guardia, J.,Gomez, J.: Nucleotide and amino acid complexity of hepatitis C virus quasispecies in serum and liver. J Virol **74**, 805-11. (2000)

Carayannopoulos, L.N.,Yokoyama, W.M.: Recognition of infected cells by natural killer cells. Curr Opin Immunol **16**, 26-33 (2004)

Chang, K.M., Rehermann, B.,Chisari, F.V.: Immunopathology of hepatitis C. Springer Semin Immunopathol **19**, 57-68 (1997)

Chang, K.M., Thimme, R., Melpolder, J.J., Oldach, D., Pemberton, J., Moorhead-Loudis, J., McHutchison, J.G., Alter, H.J.,Chisari, F.V.: Differential CD4(+) and CD8(+) T-cell responsiveness in hepatitis C virus infection. Hepatology **33**, 267-76. (2001)

Chen, W., Anton, L.C., Bennink, J.R.,Yewdell, J.W.: Dissecting the multifactorial causes of immunodominance in class I-restricted T cell responses to viruses. Immunity **12**, 83-93 (2000)

Christensen, J.E., Wodarz, D., Christensen, J.P.,Thomsen, A.R.: Perforin and IFN gamma do not significantly regulate the virus-specific CD8+ T cell response in the absence of antiviral effector activity. European Journal of Immunology **34**, 1389-1394 (2004)

Christensen, J.P., Bartholdy, C., Wodarz, D.,Thomsen, A.R.: Depletion of CD4+ T cells precipitates immunopathology in immunodeficient mice infected with a noncytocidal virus. J Immunol **166**, 3384-91. (2001)

Christensen, J.P., Cardin, R.D., Branum, K.C.,Doherty, P.C.: CD4(+) T cell-mediated control of a gamma-herpesvirus in B cell-deficient mice is mediated by IFN-gamma. Proc Natl Acad Sci U S A **96**, 5135-40 (1999)

Chun, T.W., Stuyver, L., Mizell, S.B., Ehler, L.A., Mican, J.A., Baseler, M., Lloyd, A.L., Nowak, M.A.,Fauci, A.S.: Presence of an inducible HIV-1 latent reservoir during highly active antiretroviral therapy. Proc Natl Acad Sci U S A **94**, 13193-7 (1997)

Ciurea, A., Klenerman, P., Hunziker, L., Horvath, E., Senn, B.M., Ochsenbein, A.F., Hengartner, H.,Zinkernagel, R.M.: Viral persistence in vivo through selection of neutralizing antibody-escape variants. Proc Natl Acad Sci U S A **97**, 2749-54. (2000)

Cocchi, F., DeVico, A.L., Garzino-Demo, A., Arya, S.K., Gallo, R.C., Lusso, P.: Identification of RANTES, MIP-1 alpha, and MIP-1 beta as the major HIV- suppressive factors produced by CD8+ T cells. Science **270**, 1811-5 (1995)

Condra, J.H., Schleif, W.A., Blahy, O.M., Gabryelski, L.J., Graham, D.J., Quintero, J.C., Rhodes, A., Robbins, H.L., Roth, E., Shivaprakash, M.,et al.: In vivo emergence of HIV-1 variants resistant to multiple protease inhibitors. Nature **374**, 569-71 (1995)

Cooper, S., Erickson, A.L., Adams, E.J., Kansopon, J., Weiner, A.J., Chien, D.Y., Houghton, M., Parham, P.,Walker, C.M.: Analysis of a successful immune response against hepatitis C virus. Immunity **10**, 439-49. (1999)

Curran, R., Jameson, C.L., Craggs, J.K., Grabowska, A.M., Thomson, B.J., Robins, A., Irving, W.L.,Ball, J.K.: Evolutionary trends of the first hypervariable region of the hepatitis C virus E2 protein in individuals with differing liver disease severity. J Gen Virol **83**, 11-23. (2002)

Davenport, M.P.: Benefits of memory. Nat Immunol **1**, 451. (2000)

Day, C.L., Lauer, G.M., Robbins, G.K., McGovern, B., Wurcel, A.G., Gandhi, R.T., Chung, R.T.,Walker, B.D.: Broad specificity of virus-specific CD4+ T-helper-cell responses in resolved hepatitis C virus infection. J Virol **76**, 12584-95 (2002) de Araujo, E.S., Cavalheiro Nde, P., Cubero Leitao, R.M., Borges Tosta, R.A.,Barone, A.A.: Hepatitis C viral load does not predict disease outcome: going beyond numbers. Rev Inst Med Trop Sao Paulo **44**, 71-8. (2002)

de Boer, R.J.,Boerlijst, M.C.: Diversity and virulence thresholds in AIDS. Proc Natl Acad Sci U S A **91**, 544-8 (1994)

De Boer, R.J.,Perelson, A.S.: T cell repertoires and competitive exclusion. J Theor Biol **169**, 375-90 (1994)

De Boer, R.J.,Perelson, A.S.: Towards a general function describing T cell proliferation. J Theor Biol **175**, 567-76. (1995)

De Boer, R.J.,Perelson, A.S.: Target cell limited and immune control models of HIV infection: a comparison. J Theor Biol **190**, 201-14 (1998)

Deeths, M.J., Kedl, R.M.,Mescher, M.F.: CD8+ T cells become nonresponsive (anergic) following activation in the presence of costimulation. J Immunol **163**, 102-10 (1999)

Doherty, P.C., Topham, D.J.,Tripp, R.A.: Establishment and persistence of virus-specific CD4+ and CD8+ T cell memory. Immunol Rev **150**, 23-44 (1996)

Doherty, P.C., Topham, D.J., Tripp, R.A., Cardin, R.D., Brooks, J.W., Stevenson, P.G.: Effector CD4+ and CD8+ T-cell mechanisms in the control of respiratory virus infections. Immunol Rev **159**, 105-17 (1997)

Dokun, A.O., Kim, S., Smith, H.R., Kang, H.S., Chu, D.T.,Yokoyama, W.M.: Specific and nonspecific NK cell activation during virus infection. Nat Immunol **2**, 951-6 (2001)

Douek, D.C., Brenchley, J.M., Betts, M.R., Ambrozak, D.R., Hill, B.J., Okamoto, Y., Casazza, J.P., Kuruppu, J., Kunstman, K., Wolinsky, S., Grossman, Z., Dybul, M., Oxenius, A., Price, D.A., Connors, M.,Koup, R.A.: HIV preferentially infects HIV-specific CD4+ T cells. Nature **417**, 95-8. (2002)

Effros, R.B.,Walford, R.L.: Diminished T-cell response to influenza virus in aged mice. Immunology **49**, 387-92 (1983)

Ehl, S., Klenerman, P., Aichele, P., Hengartner, H.,Zinkernagel, R.M.: A functional and kinetic comparison of antiviral effector and memory cytotoxic T lymphocyte populations in vivo and in vitro. Eur J Immunol **27**, 3404-13 (1997)

Eichelberger, M., Allan, W., Zijlstra, M., Jaenisch, R.,Doherty, P.C.: Clearance of influenza virus respiratory infection in mice lacking class I major histocompatibility complex-restricted CD8+ T cells. J Exp Med **174**, 875-80 (1991)

Fagiolo, U., Amadori, A., Cozzi, E., Bendo, R., Lama, M., Douglas, A.,Palu, G.: Humoral and cellular immune response to influenza virus vaccination in aged humans. Aging (Milano) **5**, 451-8 (1993)

Farci, P.: Hepatitis C virus. The importance of viral heterogeneity. Clin Liver Dis **5**, 895-916. (2001)

Farci, P., Shimoda, A., Coiana, A., Diaz, G., Peddis, G., Melpolder, J.C., Strazzera, A., Chien, D.Y., Munoz, S.J., Balestrieri, A., Purcell, R.H.,Alter, H.J.: The outcome of acute hepatitis C predicted by the evolution of the viral quasispecies. Science **288**, 339-44 (2000)

Finzi, D., Hermankova, M., Pierson, T., Carruth, L.M., Buck, C., Chaisson, R.E., Quinn, T.C., Chadwick, K., Margolick, J., Brookmeyer, R., Gallant, J., Markowitz, M., Ho, D.D., Richman, D.D.,Siliciano, R.F.: Identification of a reservoir for HIV-1 in patients on highly active antiretroviral therapy. Science **278**, 1295-300 (1997)

Fishman, M.A.,Perelson, A.S.: Modeling T cell-antigen presenting cell interactions. J Theor Biol **160**, 311-42. (1993)

Flemington, E.K.: Herpesvirus lytic replication and the cell cycle: arresting new developments. J Virol **75**, 4475-81 (2001)

Forns, X., Purcell, R.H.,Bukh, J.: Quasispecies in viral persistence and pathogenesis of hepatitis C virus. Trends Microbiol **7**, 402-10. (1999)

Freitas, A.A.,Rocha, B.: Population biology of lymphocytes: the flight for survival. Annu Rev Immunol **18**, 83-111 (2000)

French, A.R.,Yokoyama, W.M.: Natural killer cells and viral infections. Curr Opin Immunol **15**, 45-51 (2003)

Frost, S.D.,McLean, A.R.: Quasispecies dynamics and the emergence of drug resistance during zidovudine therapy of HIV infection. Aids **8**, 323-32 (1994)

Fujinami, R.S.: Viruses and autoimmune disease–two sides of the same coin? Trends Microbiol **9**, 377-81. (2001)

Gallo, R.C., Lusso, P.: Chemokines and HIV infection. Current Opinion of Infectious Diseases **10**, 12-17 (1997)

Garcia, F., Plana, M., Ortiz, G.M., Bonhoeffer, S., Soriano, A., Vidal, C., Cruceta, A., Arnedo, M., Gil, C., Pantaleo, G., Pumarola, T., Gallart, T., Nixon, D.F., Miro, J.M.,Gatell, J.M.: The virological and immunological consequences of structured treatment interruptions in chronic HIV-1 infection. Aids **15**, F29-40. (2001)

Gerhard, W., Mozdzanowska, K., Furchner, M., Washko, G.,Maiese, K.: Role of the B-cell response in recovery of mice from primary influenza virus infection. Immunol Rev **159**, 95-103 (1997)

Gern, J.E., Joseph, B., Galagan, D.M., Borcherding, W.R.,Dick, E.C.: Rhinovirus inhibits antigen-specific T cell proliferation through an intercellular adhesion molecule-1-dependent mechanism. J Infect Dis **174**, 1143-50. (1996)

Goon, P.K., Biancardi, A., Fast, N., Igakura, T., Hanon, E., Mosley, A.J., Asquith, B., Gould, K.G., Marshall, S., Taylor, G.P., Bangham, C.R.: Human T cell lymphotropic virus (HTLV) type-1-specific CD8+ T cells: frequency and immunodominance hierarchy. J Infect Dis **189**, 2294-8 (2004)

Goulder, P.J., Phillips, R.E., Colbert, R.A., McAdam, S., Ogg, G., Nowak, M.A., Giangrande, P., Luzzi, G., Morgan, B., Edwards, A., McMichael, A.J.,Rowland-Jones, S.: Late escape from an immunodominant cytotoxic

T-lymphocyte response associated with progression to AIDS. Nat Med **3**, 212-7 (1997a)

Goulder, P.J., Sewell, A.K., Lalloo, D.G., Price, D.A., Whelan, J.A., Evans, J., Taylor, G.P., Luzzi, G., Giangrande, P., Phillips, R.E.,McMichael, A.J.: Patterns of immunodominance in HIV-1-specific cytotoxic T lymphocyte responses in two human histocompatibility leukocyte antigens (HLA)-identical siblings with HLA-A*0201 are influenced by epitope mutation. J Exp Med **185**, 1423-33 (1997b)

Goulder, P.J.R., Phillips, R.E., Colbert, R.A., McAdam, S., Ogg, G., Nowak, M.A., Giangrande, P., Luzzi, G., Morgan, B., Edwards, A., McMichael, A.J.,RowlandJones, S.: Late escape from an immunodominant cytotoxic T-lymphocyte response associated with progression to AIDS. Nature Medicine **3**, 212-217 (1997c)

Gray, D.,Matzinger, P.: T cell memory is short-lived in the absence of antigen. J Exp Med **174**, 969-74 (1991)

Groarke, J.M.,Pevear, D.C.: Attenuated virulence of pleconaril-resistant coxsackievirus B3 variants. J Infect Dis **179**, 1538-41. (1999)

Guidotti, L.G., Ando, K., Hobbs, M.V., Ishikawa, T., Runkel, L., Schreiber, R.D.,Chisari, F.V.: Cytotoxic T lymphocytes inhibit hepatitis B virus gene expression by a noncytolytic mechanism in transgenic mice. Proc Natl Acad Sci U S A **91**, 3764-8 (1994a)

Guidotti, L.G., Ando, K., Hobbs, M.V., Ishikawa, T., Runkel, L., Schreiber, R.D.,Chisari, F.V.: Cytotoxic T-Lymphocytes Inhibit Hepatitis-B Virus Gene-Expression By a Noncytolytic Mechanism in Transgenic Mice. Proceedings of the National Academy of Sciences of the United States of America **91**, 3764-3768 (1994b)

Guidotti, L.G., Borrow, P., Brown, A., McClary, H., Koch, R.,Chisari, F.V.: Noncytopathic clearance of lymphocytic choriomeningitis virus from the hepatocyte. J Exp Med **189**, 1555-64 (1999a)

Guidotti, L.G.,Chisari, F.V.: To Kill or to Cure - Options in Host-Defense Against Viral-Infection. Current Opinion in Immunology **8**, 478-483 (1996)

Guidotti, L.G., Ishikawa, T., Hobbs, M.V., Matzke, B., Schreiber, R.,Chisari, F.V.: Intracellular inactivation of the hepatitis B virus by cytotoxic T lymphocytes. Immunity **4**, 25-36 (1996a)

Guidotti, L.G., Ishikawa, T., Hobbs, M.V., Matzke, B., Schreiber, R.,Chisari, F.V.: Intracellular Inactivation of the Hepatitis-B Virus By Cytotoxic T-Lymphocytes. Immunity **4**, 25-36 (1996b)

Guidotti, L.G., Rochford, R., Chung, J., Shapiro, M., Purcell, R.,Chisari, F.V.: Viral clearance without destruction of infected cells during acute HBV infection. Science **284**, 825-9 (1999b)

Gupta, S.,Anderson, R.M.: Population structure of pathogens: the role of immune selection. Parasitol Today **15**, 497-501 (1999)

Harrer, T., Harrer, E., Kalams, S.A., Barbosa, P., Trocha, A., Johnson, R.P., Elbeik, T., Feinberg, M.B., Buchbinder, S.P.,Walker, B.D.: Cytotoxic T lymphocytes in asymptomatic long-term nonprogressing HIV-1 infection.

Breadth and specificity of the response and relation to in vivo viral quasispecies in a person with prolonged infection and low viral load. J Immunol **156**, 2616-23 (1996a)

Harrer, T., Harrer, E., Kalams, S.A., Elbeik, T., Staprans, S.I., Feinberg, M.B., Cao, Y., Ho, D.D., Yilma, T., Caliendo, A.M., Johnson, R.P., Buchbinder, S.P.,Walker, B.D.: Strong cytotoxic T cell and weak neutralizing antibody responses in a subset of persons with stable nonprogressing HIV type 1 infection. AIDS Res Hum Retroviruses **12**, 585-92 (1996b)

Heinzinger, N.K., Bukinsky, M.I., Haggerty, S.A., Ragland, A.M., Kewalramani, V., Lee, M.A., Gendelman, H.E., Ratner, L., Stevenson, M.,Emerman, M.: The Vpr protein of human immunodeficiency virus type 1 influences nuclear localization of viral nucleic acids in nondividing host cells. Proc Natl Acad Sci U S A **91**, 7311-5 (1994)

Heise, M.T., Connick, M.,Virgin, H.W.t.: Murine cytomegalovirus inhibits interferon gamma-induced antigen presentation to CD4 T cells by macrophages via regulation of expression of major histocompatibility complex class II-associated genes. J Exp Med **187**, 1037-46 (1998a)

Heise, M.T., Pollock, J.L., O'Guin, A., Barkon, M.L., Bormley, S.,Virgin, H.W.t.: Murine cytomegalovirus infection inhibits IFN gamma-induced MHC class II expression on macrophages: the role of type I interferon. Virology **241**, 331-44 (1998b)

Hendley, J.O.: Clinical virology of rhinoviruses. Adv Virus Res **54**, 453-66 (1999)

Ho, D.D., Neumann, A.U., Perelson, A.S., Chen, W., Leonard, J.M., Markowitz, M.: Rapid Turnover of Plasma Virions and Cd4 Lymphocytes in Hiv-1 Infection. Nature **373**, 123-126 (1995)

Hoffmann, G.W.: A theory of regulation and self-non-self discrumunation in an immune network. European Journal of Immunology **5**, 638-657 (1975)

Holtappels, R., Podlech, J., Pahl-Seibert, M.F., Julch, M., Thomas, D., Simon, C.O., Wagner, M.,Reddehase, M.J.: Cytomegalovirus misleads its host by priming of CD8 T cells specific for an epitope not presented in infected tissues. J Exp Med **199**, 131-6 (2004)

Hou, S., Hyland, L., Ryan, K.W., Portner, A.,Doherty, P.C.: Virus-specific CD8+ T-cell memory determined by clonal burst size. Nature **369**, 652-4 (1994)

Hummel, M.,Abecassis, M.M.: A model for reactivation of CMV from latency. J Clin Virol **25 Suppl 2**, S123-36 (2002)

Jamieson, B.D.,Ahmed, R.: T cell memory. Long-term persistence of virus-specific cytotoxic T cells. J Exp Med **169**, 1993-2005 (1989)

Janssen, E.M., Lemmens, E.E., Wolfe, T., Christen, U., von Herrath, M.G., Schoenberger, S.P.: CD4+ T cells are required for secondary expansion and memory in CD8+ T lymphocytes. Nature **421**, 852-6 (2003)

Jeffery, K.J., Usuku, K., Hall, S.E., Matsumoto, W., Taylor, G.P., Procter, J., Bunce, M., Ogg, G.S., Welsh, K.I., Weber, J.N., Lloyd, A.L., Nowak, M.A., Nagai, M., Kodama, D., Izumo, S., Osame, M.,Bangham, C.R.: HLA alleles determine human T-lymphotropic virus-I (HTLV-I) proviral load and the risk of HTLV-I-associated myelopathy. Proc Natl Acad Sci U S A **96**, 3848-53 (1999)

Jerne, N.K.: Clonal selection in a lymphocyte network. Soc Gen Physiol Ser **29**, 39-48 (1974a)

Jerne, N.K.: Towards a network theory of the immune system. Ann Immunol (Paris) **125C**, 373-89 (1974b)

Kaech, S.M.,Ahmed, R.: Memory CD8+ T cell differentiation: initial antigen encounter triggers a developmental program in naive cells. Nat Immunol **2**, 415-22. (2001)

Kaech, S.M., Hemby, S., Kersh, E.,Ahmed, R.: Molecular and functional profiling of memory CD8 T cell differentiation. Cell **111**, 837-51 (2002)

Kagi, D., Hengartner, H.: Different Roles For Cytotoxic T-Cells in the Control of Infections With Cytopathic Versus Noncytopathic Viruses. Current Opinion in Immunology **8**, 472-477 (1996)

Kagi, D., Ledermann, B., Burki, K., Zinkernagel, R.M., Hengartner, H.: Lymphocyte- Mediated Cytotoxicity in-Vitro and in-Vivo - Mechanisms and Significance. Immunological Reviews **146**, 95-115 (1995a)

Kagi, D., Ledermann, B., Burki, K., Zinkernagel, R.M.,Hengartner, H.: Molecular Mechanisms of Lymphocyte-Mediated Cytotoxicity and Their Role in Immunological Protection and Pathogenesis in-Vivo. Annual Review of Immunology **14**, 207-232 (1996)

Kagi, D., Seiler, P., Pavlovic, J., Ledermann, B., Burki, K., Zinkernagel, R.M.,Hengartner, H.: The Roles of Perforin-Dependent and Fas-Dependent Cytotoxicity in Protection Against Cytopathic and Noncytopathic Viruses. European Journal of Immunology **25**, 3256-3262 (1995b)

Kalams, S.A., Goulder, P.J., Shea, A.K., Jones, N.G., Trocha, A.K., Ogg, G.S.,Walker, B.D.: Levels of Human Immunodeficiency Virus Type 1-Specific Cytotoxic T- Lymphocyte Effector and Memory Responses Decline after Suppression of Viremia with Highly Active Antiretroviral Therapy. J Virol **73**, 6721-6728 (1999)

Karrer, U., Sierro, S., Wagner, M., Oxenius, A., Hengel, H., Koszinowski, U.H., Phillips, R.E., Klenerman, P.: Memory inflation: continuous accumulation of antiviral CD8+ T cells over time. J Immunol **170**, 2022-9 (2003)

Kaufmann, D.E., Lichterfeld, M., Altfeld, M., Addo, M.M., Johnston, M.N., Lee, P.K., Wagner, B.S., Kalife, E.T., Strick, D., Rosenberg, E.S.,Walker, B.D.: Limited durability of viral control following treated acute HIV infection. PLoS Med **1**, e36 (2004)

Kaur, A., Daniel, M.D., Hempel, D., Lee-Parritz, D., Hirsch, M.S., Johnson, R P · Cytotoxic T-lymphocyte responses to cytomegalo virus in normal and simian immunodeficiency virus-infected rhesus macaques. J Virol **70**, 7725-33 (1996)

King, C.C., de Fries, R., Kolhekar, S.R.,Ahmed, R.: In vivo selection of lymphocyte-tropic and macrophage-tropic variants of lymphocytic choriomeningitis virus during persistent infection. J Virol **64**, 5611-6 (1990)

Klavinskis, L.S., Geckeler, R.,Oldstone, M.B.: Cytotoxic T lymphocyte control of acute lymphocytic choriomeningitis virus infection: interferon gamma, but not tumour necrosis factor alpha, displays antiviral activity in vivo. J Gen Virol **70**, 3317-25. (1989)

Klenerman, P., Lechner, F., Kantzanou, M., Ciurea, A., Hengartner, H., Zinkernagel, R.: Viral escape and the failure of cellular immune responses. Science **289**, 2003. (2000)

Klenerman, P., Meier, U.C., Phillips, R.E.,McMichael, A.J.: The effects of natural altered peptide ligands on the whole blood cytotoxic T lymphocyte response to human immunodeficiency virus. Eur J Immunol **25**, 1927-31 (1995)

Klenerman, P., Phillips, R.E., Rinaldo, C.R., Wahl, L.M., Ogg, G., May, R.M., McMichael, A.J.,Nowak, M.A.: Cytotoxic T lymphocytes and viral turnover in HIV type 1 infection. Proc Natl Acad Sci U S A **93**, 15323-8 (1996)

Komarova, N.L., Barnes, E., Klenerman, P.,Wodarz, D.: Boosting immunity by antiviral drug therapy: a simple relationship among timing, efficacy, and success. Proc Natl Acad Sci U S A **100**, 1855-60 (2003)

Komatsu, H., Sierro, S., A, V.C.,Klenerman, P.: Population analysis of antiviral T cell responses using MHC class I-peptide tetramers. Clin Exp Immunol **134**, 9-12 (2003)

Krakauer, D.C.,Nowak, M.: T-cell induced pathogenesis in HIV: bystander effects and latent infection. Proc R Soc Lond B Biol Sci **266**, 1069-75 (1999)

Kundig, T.M., Bachmann, M.F., Oehen, S., Hoffmann, U.W., Simard, J.J., Kalberer, C.P., Pircher, H., Ohashi, P.S., Hengartner, H., Zinkernagel, R.M.: On the role of antigen in maintaining cytotoxic T-cell memory. Proc Natl Acad Sci U S A **93**, 9716-23 (1996a)

Kundig, T.M., Bachmann, M.F., Ohashi, P.S., Pircher, H., Hengartner, H., Zinkernagel, R.M.: On T cell memory: arguments for antigen dependence. Immunol Rev **150**, 63-90 (1996b)

Larder, B.A., Darby, G.,Richman, D.D.: HIV with reduced sensitivity to zidovudine (AZT) isolated during prolonged therapy. Science **243**, 1731-4 (1989)

Lau, L.L., Jamieson, B.D., Somasundaram, T.,Ahmed, R.: Cytotoxic T-cell memory without antigen. Nature **369**, 648-52 (1994)

Layden, T.J., Lam, N.P.,Wiley, T.E.: Hepatitis C viral dynamics. Clin Liver Dis **3**, 793-810. (1999)

Layden, T.J., Mika, B.,Wiley, T.E.: Hepatitis C kinetics: mathematical modeling of viral response to therapy. Semin Liver Dis **20**, 173-83 (2000)

Lechner, F., Gruener, N.H., Urbani, S., Uggeri, J., Santantonio, T., Kammer, A.R., Cerny, A., Phillips, R., Ferrari, C., Pape, G.R., Klenerman, P.: CD8+ T lymphocyte responses are induced during acute hepatitis C virus infection but are not sustained. Eur J Immunol **30**, 2479-87. (2000a)

Lechner, F., Sullivan, J., Spiegel, H., Nixon, D.F., Ferrari, B., Davis, A., Borkowsky, B., Pollack, H., Barnes, E., Dusheiko, G.,Klenerman, P.: Why do cytotoxic T lymphocytes fail to eliminate hepatitis C virus? Lessons from studies using major histocompatibility complex class I peptide tetramers. Philos Trans R Soc Lond B Biol Sci **355**, 1085-92. (2000b)

Lechner, F., Wong, D.K., Dunbar, P.R., Chapman, R., Chung, R.T., Dohrenwend, P., Robbins, G., Phillips, R., Klenerman, P.,Walker, B.D.: Analysis of successful immune responses in persons infected with hepatitis C virus. J Exp Med **191**, 1499-512 (2000c)

Lehmann-Grube, F.: Lymphocytic choriomeningitis virus. Virol. Monogr. **10**, 1-173 (1971)

Levin, S.A., Grenfell, B., Hastings, A.,Perelson, A.S.: Mathematical and computational challenges in population biology and ecosystems science. Science **275**, 334-343 (1997)

Levy, J.A., Mackewicz, C.E.,Barker, E.: Controlling HIV pathogenesis: the role of the noncytotoxic anti-HIV response of CD8+ T cells. Immunol Today **17**, 217-24 (1996)

Lifson, J.D., Rossio, J.L., Arnaout, R., Li, L., Parks, T.L., Schneider, D.M., Kiser, R.F., Coalter, V.J., Walsh, G., Imming, R., Fischer, B., Flynn, B.M., Nowak, M.A.,Wodarz, D.: Containment of SIV infection: cellular immune responses and protection from rechallenge following transient post-inoculation antiretroviral treatment. J. Virol. **74**, 2584-93 (2000)

Lifson, J.D., Rossio, J.L., Piatak, M., Jr., Parks, T., Li, L., Kiser, R., Coalter, V., Fisher, B., Flynn, B.M., Czajak, S., Hirsch, V.M., Reimann, K.A., Schmitz, J.E., Ghrayeb, J., Bischofberger, N., Nowak, M.A., Desrosiers, R.C.,Wodarz, D.: Role of CD8(+) lymphocytes in control of simian immunodeficiency virus infection and resistance to rechallenge after transient early antiretroviral treatment. J Virol **75**, 10187-99. (2001)

Lin, J., Andreasen, V.,Levin, S.A.: Dynamics of influenza A drift: the linear three-strain model. Math Biosci **162**, 33-51. (1999)

Lipsitch, M.,Moxon, E.R.: Virulence and transmissibility of pathogens: what is the relationship? Trends Microbiol **5**, 31-7. (1997)

Lisziewicz, J.,Lori, F.: Structured treatment interruptions in HIV/AIDS therapy. Microbes Infect **4**, 207-14. (2002)

Little, S.J., McLean, A.R., Spina, C.A., Richman, D.D.,Havlir, D.V.: Viral dynamics of acute HIV-1 infection. J Exp Med **190**, 841-50. (1999)

Liu, H., Andreansky, S., Diaz, G., Turner, S.J., Wodarz, D.,Doherty, P.C.: Quantitative analysis of long-term virus-specific CD8+-T-cell memory in mice challenged with unrelated pathogens. J Virol **77**, 7756-63 (2003)

Lori, F., Lewis, M.G., Xu, J., Varga, G., Zinn, D.E., Jr., Crabbs, C., Wagner, W., Greenhouse, J., Silvera, P., Yalley-Ogunro, J., Tinelli, C.,Lisziewicz, J.: Control of SIV rebound through structured treatment interruptions during early infection. Science **290**, 1591-3. (2000a)

Lori, F., Maserati, R., Foli, A., Seminari, E., Timpone, J.,Lisziewicz, J.: Structured treatment interruptions to control HIV-1 infection. Lancet **355**, 287-8. (2000b)

Lotka, A.J. Elements of Mathematical Biology. Dover, New York: Dover Pubns(1956)

Lyra, A.C., Fan, X., Lang, D.M., Yusim, K., Ramrakhiani, S., Brunt, E.M., Korber, B., Perelson, A.S.,Di Bisceglie, A.M.: Evolution of hepatitis C viral quasispecies after liver transplantation. Gastroenterology **123**, 1485-93. (2002)

Major, M.E., Mihalik, K., Fernandez, J., Seidman, J., Kleiner, D., Kolykhalov, A.A., Rice, C.M.,Feinstone, S.M.: Long-term follow-up of chimpanzees inoculated with the first infectious clone for hepatitis C virus. J Virol **73**, 3317-25. (1999)

Manjunath, N., Shankar, P., Wan, J., Weninger, W., Crowley, M.A., Hieshima, K., Springer, T.A., Fan, X., Shen, H., Lieberman, J.,von Andrian, U.H.: Effector differentiation is not prerequisite for generation of memory cytotoxic T lymphocytes. J Clin Invest **108**, 871-8 (2001)

Manzin, A., Solforosi, L., Giostra, F., Bianchi, F.B., Bruno, S., Rossi, S., Gabrielli, A., Candela, M., Petrelli, E.,Clementi, M.: Quantitative analysis of hepatitis C virus activity in vivo in different groups of untreated patients. Arch Virol **142**, 465-72 (1997)

Matloubian, M., Suresh, M., Glass, A., Galvan, M., Chow, K., Whitmire, J.K., Walsh, C.M., Clark, W.R.,Ahmed, R.: A role for perforin in downregulating T-cell responses during chronic viral infection. J Virol **73**, 2527-36 (1999)

Matzinger, P.: An innate sense of danger. Semin Immunol **10**, 399-415. (1998)

McAdam, S., Klenerman, P., Tussey, L., Rowland-Jones, S., Lalloo, D., Phillips, R., Edwards, A., Giangrande, P., Brown, A.L., Gotch, F.,et al.: Immunogenic HIV variant peptides that bind to HLA-B8 can fail to stimulate cytotoxic T lymphocyte responses. J Immunol **155**, 2729-36 (1995)

McMichael, A., Goulder, P., Rowlandjones, S., Nowak, M.,Phillips, R.: Hiv Escapes From Cytoxic Lymphocytes. Immunology **89**, SE111-SE111 (1996)

McMichael, A., Rowlandjones, S., Klenerman, P., McAdam, S., Gotch, F., Phillips, R.,Nowak, M.: Epitope Variation and T-Cell Recognition. Journal of Cellular Biochemistry , **60-60** (1995)

McMichael, A.J.,Phillips, R.E.: Escape of human immunodeficiency virus from immune control. Annual Review of Immunology **15**, 271-296 (1997)

McMichael, A.J.,Rowland-Jones, S.L.: Cellular immune responses to HIV. Nature **410**, 980-7. (2001)

Mercado, R., Vijh, S., Allen, S.E., Kerksiek, K., Pilip, I.M.,Pamer, E.G.: Early programming of T cell populations responding to bacterial infection. J Immunol **165**, 6833-9 (2000)

Miller, V.: Structured treatment interruptions in antiretroviral management of HIV- 1. Curr Opin Infect Dis **14**, 29-37. (2001)

Montaner, L.J.: Structured treatment interruptions to control HIV-1 and limit drug exposure. Trends Immunol **22**, 92-6. (2001)

Moore, J.P., Trkola, A.,Dragic, T.: Co-receptors for HIV-1 entry. Curr Opin Immunol **9**, 551-62 (1997)

Moskophidis, D., Battegay, M., van den Broek, M., Laine, E., Hoffmann-Rohrer, U.,Zinkernagel, R.M.: Role of virus and host variables in virus persistence or immunopathological disease caused by a non-cytolytic virus. J Gen Virol **76**, 381-91 (1995a)

Moskophidis, D., Battegay, M., van den Broek, M., Laine, E., Hoffmann-Rohrer, U.,Zinkernagel, R.M.: Role of virus and host variables in virus persistence or immunopathological disease caused by a non-cytolytic virus. J Gen Virol **76**, 381-91. (1995b)

Moskophidis, D.,Kioussis, D.: Contribution of virus-specific CD8+ cytotoxic T cells to virus clearance or pathologic manifestations of influenza virus infection in a T cell receptor transgenic mouse model. J Exp Med **188**, 223-32. (1998)

Moskophidis, D., Laine, E.,Zinkernagel, R.M.: Peripheral clonal deletion of antiviral memory CD8+ T cells. Eur J Immunol **23**, 3306-11. (1993a)

Moskophidis, D., Lechner, F., Pircher, H.,Zinkernagel, R.M.: Virus persistence in acutely infected immunocompetent mice by exhaustion of antiviral cytotoxic effector T cells. Nature **362**, 758-61. (1993b)

Moskophidis, D., Lechner, F., Pircher, H.,Zinkernagel, R.M.: Virus persistence in acutely infected immunocompetent mice by exhaustion of antiviral cytotoxic effector T cells [published erratum appears in Nature 1993 Jul 15;364(6434):262]. Nature **362**, 758-61 (1993c)

Mullbacher, A.: The long-term maintenance of cytotoxic T cell memory does not require persistence of antigen. J Exp Med **179**, 317-21 (1994)

Murali-Krishna, K., Lau, L.L., Sambhara, S., Lemonnier, F., Altman, J., Ahmed, R.: Persistence of memory CD8 T cells in MHC class I-deficient mice. Science **286**, 1377-81 (1999)

Nansen, A., Jensen, T., Christensen, J.P., Andreasen, S.O., Ropke, C., Marker, O.,Thomsen, A.R.: Compromised virus control and augmented perforin-mediated immunopathology in IFN-gamma-deficient mice infected with lymphocytic choriomeningitis virus. J Immunol **163**, 6114-22. (1999)

Neumann, A.U., Lam, N.P., Dahari, H., Gretch, D.R., Wiley, T.E., Layden, T.J.,Perelson, A.S.: Hepatitis C viral dynamics in vivo and the antiviral efficacy of interferon-alpha therapy. Science **282**, 103-7. (1998)

Northfield, J., Lucas, M., Jones, H., Young, N.T.,Klenerman, P.: Does memory improve with age? CD85j (ILT-2/LIR-1) expression on CD8 T cells correlates with 'memory inflation' in human cytomegalovirus infection. Immunol Cell Biol **83**, 182-8 (2005)

Nowak, M.: Hiv Mutation-Rate. Nature **347**, 522-522 (1990)

Nowak, M.A.: Variability of Hiv Infections. Journal of Theoretical Biology **155**, 1-20 (1992)

Nowak, M.A.: Immune-Responses Against Multiple Epitopes - a Theory For Immunodominance and Antigenic Variation. Seminars in Virology **7**, 83-92 (1996)

Nowak, M.A., Anderson, R.M., McLean, A.R., Wolfs, T.F.W., Goudsmit, J.,May, R.M.: Antigenic Diversity Thresholds and the Development of Aids. Science **254**, 963-969 (1991)

Nowak, M.A.,Bangham, C.R.: Population dynamics of immune responses to persistent viruses. Science **272**, 74-9 (1996)

Nowak, M.A., Lloyd, A.L., Vasquez, G.M., Wiltrout, T.A., Wahl, L.M., Bischofberger, N., Williams, J., Kinter, A., Fauci, A.S., Hirsch, V.M.,Lifson, J.D.: Viral dynamics of primary viremia and antiretroviral therapy in simian immunodeficiency virus infection. J Virol **71**, 7518-25 (1997)

Nowak, M.A.,May, R.M.: Mathematical Biology of Hiv Infections - Antigenic Variation and Diversity Threshold. Mathematical Biosciences **106**, 1-21 (1991)

Nowak, M.A.,May, R.M. Virus dynamics. Mathematical principles of immunology and virology.: Oxford University Press(2000)

Nowak, M.A., May, R.M., Phillips, R.E., Rowlandjones, S., Lalloo, D.G., McAdam, S., Klenerman, P., Koppe, B., Sigmund, K., Bangham, C.R.M., McMichael, A.J.: Antigenic Oscillations and Shifting Immunodominance in Hiv-1 Infections. Nature **375**, 606-611 (1995a)

Nowak, M.A., May, R.M.,Sigmund, K.: Immune-Responses Against Multiple Epitopes. Journal of Theoretical Biology **175**, 325-353 (1995b)

Ochsenbein, A.F., Karrer, U., Klenerman, P., Althage, A., Ciurea, A., Shen, H., Miller, J.F., Whitton, J.L., Hengartner, H.,Zinkernagel, R.M.: A comparison of T cell memory against the same antigen induced by virus versus intracellular bacteria. Proc Natl Acad Sci U S A **96**, 9293-8 (1999)

Oehen, S., Waldner, H., Kundig, T.M., Hengartner, H.,Zinkernagel, R.M.: Antivirally protective cytotoxic T cell memory to lymphocytic choriomeningitis virus is governed by persisting antigen. J Exp Med **176**, 1273-81 (1992)

Ogg, G.S., Jin, X., Bonhoeffer, S., Moss, P., Nowak, M.A., Monard, S., Segal, J.P., Cao, Y., Rowland-Jones, S.L., Hurley, A., Markowitz, M., Ho, D.D., McMichael, A.J.,Nixon, D.F.: Decay kinetics of human immunodeficiency virus-specific effector cytotoxic T lymphocytes after combination antiretroviral therapy. J Virol **73**, 797-800 (1999)

Oldstone, M.B., Salvato, M., Tishon, A.,Lewicki, H.: Virus-lymphocyte interactions. III. Biologic parameters of a virus variant that fails to generate CTL and establishes persistent infection in immunocompetent hosts. Virology **164**, 507-16 (1988)

Oldstone, M.B., Sinha, Y.N., Blount, P., Tishon, A., Rodriguez, M., von Wedel, R.,Lampert, P.W.: Virus-induced alterations in homeostasis: alteration in differentiated functions of infected cells in vivo. Science **218**, 1125-7. (1982)

Ortiz, G.M., Hu, J., Goldwitz, J.A., Chandwani, R., Larsson, M., Bhardwaj, N., Bonhoeffer, S., Ramratnam, B., Zhang, L., Marko- witz, M.M., Nixon, D.F.: Residual viral replication during antiretroviral therapy boosts human immunodeficiency virus type 1-specific CD8+ T-cell responses in subjects treated early after infection. J Virol **76**, 411-5. (2002)

Ortiz, G.M., Nixon, D.F., Trkola, A., Binley, J., Jin, X., Bonhoeffer, S., Kuebler, P.J., Donahoe, S.M., Demoitie, M.A., Kakimoto, W.M., Ketas, T., Clas, B., Heymann, J.J., Zhang, L., Cao, Y., Hurley, A., Moore, J.P., Ho, D.D.,Markowitz, M.: HIV-1-specific immune responses in subjects who temporarily contain virus replication after discontinuation of highly active antiretroviral therapy. J Clin Invest **104**, R13-8 (1999)

Ortiz, G.M., Wellons, M., Brancato, J., Vo, H.T., Zinn, R.L., Clarkson, D.E., Van Loon, K., Bonhoeffer, S., Miralles, G.D., Montefiori, D., Bartlett, J.A.,Nixon, D.F.: Structured antiretroviral treatment interruptions in chronically HIV-1- infected subjects. Proc Natl Acad Sci U S A **98**, 13288-93. (2001)

Oxenius, A., Price, D.A., Gunthard, H.F., Dawson, S.J., Fagard, C., Perrin, L., Fischer, M., Weber, R., Plana, M., Garcia, F., Hirschel, B., McLean, A.,Phillips, R.E.: Stimulation of HIV-specific cellular immunity by structured treatment interruption fails to enhance viral control in chronic HIV infection. Proc Natl Acad Sci U S A **99**, 13747-52 (2002)

Perelson, A.S.: Modelling viral and immune system dynamics. Nature Rev Immunol **2**, 28-36. (2002)

Perelson, A.S., Essunger, P., Cao, Y., Vesanen, M., Hurley, A., Saksela, K., Markowitz, M.,Ho, D.D.: Decay characteristics of HIV-1-infected compartments during combination therapy. Nature **387**, 188-91. (1997)

Perelson, A.S., Neumann, A.U., Markowitz, M., Leonard, J.M.,Ho, D.D.: Hiv-1 Dynamics in-Vivo - Virion Clearance Rate, Infected Cell Life- Span, and Viral Generation Time. Science **271**, 1582-1586 (1996)

Pevear, D.C., Tull, T.M., Seipel, M.E.,Groarke, J.M.: Activity of pleconaril against enteroviruses. Antimicrob Agents Chemother **43**, 2109-15. (1999)

Phillips, R.E., Rowland-Jones, S., Nixon, D.F., Gotch, F.M., Edwards, J.P., Ogunlesi, A.O., Elvin, J.G., Rothbard, J.A., Bangham, C.R., Rizza, C.R.,et al.: Human immunodeficiency virus genetic variation that can escape cytotoxic T cell recognition. Nature **354**, 453-9 (1991)

hskip -1.0cm Pier, G.B., Lyczak, J.B.,Wetzler, L.M. Immunology, Infection, and Immunity. Washington D.C.: ASM Press(2004)

Planz, O., Ehl, S., Furrer, E., Horvath, E., Brundler, M.A., Hengartner, H.,Zinkernagel, R.M.: A critical role for neutralizing-antibody-producing B cells, CD4(+) T cells, and interferons in persistent and acute infections of mice with lymphocytic choriomeningitis virus: implications for adoptive immunotherapy of virus carriers. Proc Natl Acad Sci U S A **94**, 6874-9 (1997)

Pontisso, P., Bellati, G., Brunetto, M., Chemello, L., Colloredo, G., Di Stefano, R., Nicoletti, M., Rumi, M.G., Ruvoletto, M.G., Soffredini, R., Valenza,

L.M.,Colucci, G.: Hepatitis C virus RNA profiles in chronically infected individuals: do they relate to disease activity? Hepatology **29**, 585-9. (1999)

Price, D.A., Goulder, P.J., Klenerman, P., Sewell, A.K., Easterbrook, P.J., Troop, M., Bangham, C.R.,Phillips, R.E.: Positive selection of HIV-1 cytotoxic T lymphocyte escape variants during primary infection. Proc Natl Acad Sci U S A **94**, 1890-5 (1997a)

Price, D.A., Goulder, P.J.R., Klenerman, P., Sewell, A.K., Easterbrook, P.J., Troop, M., Bangham, C.R.M.,Phillips, R.E.: Positive selection of HIV-1 cytotoxic T lymphocyte escape variants during primary infection. Proceedings of the National Academy of Sciences of the United States of America **94**, 1890-1895 (1997b)

Price, D.A., O'Callaghan C, A., Whelan, J.A., Easterbrook, P.J.,Phillips, R.E.: Cytotoxic T lymphocytes and viral evolution in primary HIV-1 infection. Clin Sci (Colch) **97**, 707-18. (1999)

Puoti, C., Stati, T.,Magrini, A.: Serum HCV RNA titer does not predict the severity of liver damage in HCV carriers with normal aminotransferase levels. Liver **19**, 104-9. (1999)

Quinnan, G.V., Manischewitz, J.E.,Ennis, P.A.: Role of cytotoxic T lymphocytes in murine cytomegalovirus infection. J Gen Virol **47**, 503-8 (1980)

Reddehase, M.J., Podlech, J.,Grzimek, N.K.: Mouse models of cytomegalovirus latency: overview. J Clin Virol **25 Suppl 2**, S23-36 (2002)

Regoes, R.R., Wodarz, D.,Nowak, M.A.: Virus dynamics: the effect of target cell limitation and immune responses on virus evolution. J Theor Biol **191**, 451-62 (1998)

Reusser, P., Cathomas, G., Attenhofer, R., Tamm, M.,Thiel, G.: Cytomegalovirus (CMV)-specific T cell immunity after renal transplantation mediates protection from CMV disease by limiting the systemic virus load. J Infect Dis **180**, 247-53 (1999)

Ribeiro, R.M.,Bonhoeffer, S.: Production of resistant HIV mutants during antiretroviral therapy. Proc Natl Acad Sci U S A **97**, 7681-6 (2000)

Ribeiro, R.M., Bonhoeffer, S.,Nowak, M.A.: The frequency of resistant mutant virus before antiviral therapy. Aids **12**, 461-5 (1998)

Richman, D.D.: Drug resistance in viruses. Trends Microbiol **2**, 401-7 (1994)

Richman, D.D.: Antiretroviral drug resistance: mechanisms, pathogenesis, clinical significance. Adv Exp Med Biol **394**, 383-95 (1996)

Richman, D.D.: How drug resistance arises. Sci Am **279**, 88. (1998)

Ridge, J.P., Di Rosa, F.,Matzinger, P.: A conditioned dendritic cell can be a temporal bridge between a CD4+ T- helper and a T-killer cell. Nature **393**, 474-8 (1998)

Rosenberg, E.S., Altfeld, M., Poon, S.H., Phillips, M.N., Wilkes, B.M., Eldridge, R.L., Robbins, G.K., D'Aquila, R.T., Goulder, P.J., Walker, B.D.: Immune control of HIV-1 after early treatment of acute infection. Nature **407**, 523-6 (2000)

Rosenberg, E.S., Billingsley, J.M., Caliendo, A.M., Boswell, S.L., Sax, P.E., Kalams, S.A.,Walker, B.D.: Vigorous HIV-1-specific CD4+ T cell responses associated with control of viremia. Science **278**, 1447-50 (1997)

Rosenberg, E.S., LaRosa, L., Flynn, T., Robbins, G.,Walker, B.D.: Characterization of HIV-1-specific T-helper cells in acute and chronic infection. Immunol Lett **66**, 89-93 (1999)

Rosenberg, E.S.,Walker, B.D.: HIV type 1-specific helper T cells: a critical host defense. AIDS Res Hum Retroviruses **14 Suppl 2**, S143-7 (1998)

Rouse, B.T.,Deshpande, S.: Viruses and autoimmunity: an affair but not a marriage contract. Rev Med Virol **12**, 107-13. (2002)

Rudensey, L.M., Kimata, J.T., Benveniste, R.E.,Overbaugh, J.: Progression to AIDS in macaques is associated with changes in the replication, tropism, and cytopathic properties of the simian immunodeficiency virus variant population. Virology **207**, 528-42 (1995)

Saag, M.S., Hahn, B.H., Gibbons, J., Li, Y.X., Parks, E.S., Parks, W.P.,Shaw, G.M.: Extensive Variation of Human Immunodeficiency Virus Type-1 In-vivo. Nature **334**, 440-444 (1988)

Saah, A.J., Hoover, D.R., Weng, S., Carrington, M., Mellors, J., Rinaldo, C.R., Jr., Mann, D., Apple, R., Phair, J.P., Detels, R., O'Brien, S., Enger, C., Johnson, P.,Kaslow, R.A.: Association of HLA profiles with early plasma viral load, CD4+ cell count and rate of progression to AIDS following acute HIV-1 infection. Multicenter AIDS Cohort Study. Aids **12**, 2107-13 (1998)

Sallusto, F.,Lanzavecchia, A.: Exploring pathways for memory T cell generation. J Clin Invest **108**, 805-6 (2001)

Sarawar, S.R., Lee, B.J., Reiter, S.K.,Schoenberger, S.P.: Stimulation via CD40 can substitute for CD4 T cell function in preventing reactivation of a latent herpesvirus. Proc Natl Acad Sci U S A **98**, 6325-9. (2001)

Scarlatti, G., Tresoldi, E., Bjorndal, A., Fredriksson, R., Colognesi, C., Deng, H.K., Malnati, M.S., Plebani, A., Siccardi, A.G., Littman, D.R., Fenyo, E.M.,Lusso, P.: In vivo evolution of HIV-1 co-receptor usage and sensitivity to chemokine-mediated suppression. Nat Med **3**, 1259-65 (1997)

Schmitz, J.E., Kuroda, M.J., Santra, S., Sasseville, V.G., Simon, M.A., Lifton, M.A., Racz, P., Tenner-Racz, K., Dalesandro, M., Scallon, B.J., Ghrayeb, J., Forman, M.A., Montefiori, D.C., Rieber, E.P., Letvin, N.L.,Reimann, K.A.: Control of viremia in simian immunodeficiency virus infection by CD8(+) lymphocytes. Science **283**, 857-60 (1999)

Schoenberger, S.P., Toes, R.E., van der Voort, E.I., Offringa, R.,Melief, C.J.: T-cell help for cytotoxic T lymphocytes is mediated by CD40-CD40L interactions. Nature **393**, 480-3 (1998)

Scholz, M., Doerr, H.W.,Cinatl, J.: Inhibition of cytomegalovirus immediate early gene expression: a therapeutic option? Antiviral Res **49**, 129-45 (2001)

Seewaldt, S., Thomas, H.E., Ejrnaes, M., Christen, U., Wolfe, T., Rodrigo, E., Coon, B., Michelsen, B., Kay, T.W.,von Herrath, M.G.: Virus-induced autoimmune diabetes: most beta-cells die through inflammatory cytokines and not perforin from autoreactive (anti-viral) cytotoxic T-lymphocytes. Diabetes **49**, 1801-9. (2000)

Selin, L.K., Lin, M.Y., Kraemer, K.A., Pardoll, D.M., Schneck, J.P., Varga, S.M., Santolucito, P.A., Pinto, A.K.,Welsh, R.M.: Attrition of T cell memory: selective loss of LCMV epitope-specific memory CD8 T cells following infections with heterologous viruses. Immunity **11**, 733-42 (1999)

Selin, L.K., Vergilis, K., Welsh, R.M.,Nahill, S.R.: Reduction of otherwise remarkably stable virus-specific cytotoxic T lymphocyte memory by heterologous viral infections. J Exp Med **183**, 2489-99 (1996)

Shedlock, D.J.,Shen, H.: Requirement for CD4 T cell help in generating functional CD8 T cell memory. Science **300**, 337-9 (2003)

Sierro, S., Rothkopf, R.,Klenerman, P.: Evolution of diverse antiviral CD8+ T cell populations after murine cytomegalovirus infection. Eur J Immunol **35**, 1113-23 (2005)

Silvestri, G., Fedanov, A., Germon, S., Kozyr, N., Kaiser, W.J., Garber, D.A., McClure, H., Feinberg, M.B.,Staprans, S.I.: Divergent host responses during primary simian immunodeficiency virus SIVsm infection of natural sooty mangabey and nonnatural rhesus macaque hosts. J Virol **79**, 4043-54 (2005)

Slifka, M.K.,Ahmed, R.: Long-lived plasma cells: a mechanism for maintaining persistent antibody production. Curr Opin Immunol **10**, 252-8 (1998)

Stepp, S.E., Mathew, P.A., Bennett, M., de Saint Basile, G.,Kumar, V.: Perforin: more than just an effector molecule. Immunol Today **21**, 254-6. (2000)

Stevenson, M.: Identification of factors that govern HIV-1 replication in non-dividing host cells. AIDS Res Hum Retroviruses **10 Suppl 1**, S11-5 (1994)

Stevenson, M.: Pathway to understanding AIDS. Nature Structural Biology **3**, 303-306 (1996)

Stevenson, M., Brichacek, B., Heinzinger, N., Swindells, S., Pirruccello, S., Janoff, E.,Emerman, M.: Molecular basis of cell cycle dependent HIV-1 replication. Implications for control of virus burden. Adv Exp Med Biol **374**, 33-45 (1995)

Sun, J.C.,Bevan, M.J.: Defective CD8 T cell memory following acute infection without CD4 T cell help. Science **300**, 339-42 (2003)

Sun, J.C., Williams, M.A.,Bevan, M.J.: CD4+ T cells are required for the maintenance, not programming, of memory CD8+ T cells after acute infection. Nat Immunol **5**, 927-33 (2004)

Swain, S.L.: Regulation of the generation and maintenance of T-cell memory: a direct, default pathway from effectors to memory cells. Microbes Infect **5**, 213-9 (2003)

Tersmette, M., Lange, J.M., de Goede, R.E., de Wolf, F., Eeftink-Schattenkerk, J.K., Schellekens, P.T., Coutinho, R.A., Huisman, J.G., Goudsmit, J., Miedema, F.: Association between biological properties of human immunodeficiency virus variants and risk for AIDS and AIDS mortality. Lancet **1**, 983-5 (1989)

Thimme, R., Oldach, D., Chang, K.M., Steiger, C., Ray, S.C.,Chisari, F.V.: Determinants of viral clearance and persistence during acute hepatitis C virus infection. J Exp Med **194**, 1395-406. (2001)

Thomsen, A.R., Johansen, J., Marker, O.,Christensen, J.P.: Exhaustion of CTL memory and recrudescence of viremia in lymphocytic choriomeningitis virus-infected MHC class II-deficient mice and B cell- deficient mice. J Immunol **157**, 3074-80 (1996)

Thomsen, A.R.,Marker, O.: The complementary roles of cellular and humoral immunity in resistance to re-infection with LCM virus. Immunology **65**, 9-15 (1988)

Thomsen, A.R., Nansen, A., Andersen, C., Johansen, J., Marker, O.,Christensen, J.P.: Cooperation of B cells and T cells is required for survival of mice infected with vesicular stomatitis virus. Int Immunol **9**, 1757-66 (1997)

Thomsen, A.R., Nansen, A., Andreasen, S.O., Wodarz, D.,Christensen, J.P.: Host factors influencing viral persistence. Phil. Trans. Roy. Soc. Lond. B **355**, 1031-1041 (2000)

Thomsen, A.R., Nansen, A., Christensen, J.P., Andreasen, S.O.,Marker, O.: CD40 ligand is pivotal to efficient control of virus replication in mice infected with lymphocytic choriomeningitis virus. J Immunol **161**, 4583-90 (1998)

Thomson, M., Nascimbeni, M., Gonzales, S., Murthy, K.K., Rehermann, B.,Liang, T.J.: Emergence of a distinct pattern of viral mutations in chimpanzees infected with a homogeneous inoculum of hepatitis C virus. Gastroenterology **121**, 1226-33. (2001)

Thorley-Lawson, D.A.,Babcock, G.J.: A model for persistent infection with Epstein-Barr virus: the stealth virus of human B cells. Life Sci **65**, 1433-53 (1999)

Topham, D.J., Tripp, R.A.,Doherty, P.C.: CD8+ T cells clear influenza virus by perforin or Fas-dependent processes. J Immunol **159**, 5197-200 (1997)

Topham, D.J., Tripp, R.A., Hamilton-Easton, A.M., Sarawar, S.R., Doherty, P.C.: Quantitative analysis of the influenza virus-specific CD4+ T cell memory in the absence of B cells and Ig. J Immunol **157**, 2947-52. (1996)

Tripp, R.A., Sarawar, S.R.,Doherty, P.C.: Characteristics of the influenza virus-specific CD8+ T cell response in mice homozygous for disruption of the H-2lAb gene. J Immunol **155**, 2955-9 (1995)

Urbain, J.: Idiotypic networks: a noisy background or a breakthrough in immunological thinking? The broken mirror hypothesis. Ann Inst Pasteur Immunol **137C**, 57-64 (1986)

van den Broek, M.F., Muller, U., Huang, S., Aguet, M.,Zinkernagel, R.M.: Antiviral defense in mice lacking both alpha/beta and gamma interferon receptors. J Virol **69**, 4792-6. (1995a) van den Broek, M.F., Muller, U., Huang, S., Zinkernagel, R.M.,Aguet, M.: Immune defence in mice lacking type I and/or type II interferon receptors. Immunol Rev **148**, 5-18. (1995b)

van Stipdonk, M.J., Hardenberg, G., Bijker, M.S., Lemmens, E.E., Droin, N.M., Green, D.R.,Schoenberger, S.P.: Dynamic programming of CD8+ T lymphocyte responses. Nat Immunol **4**, 361-5 (2003)

van Stipdonk, M.J., Lemmens, E.E.,Schoenberger, S.P.: Naive CTLs require a single brief period of antigenic stimulation for clonal expansion and differentiation. Nat Immunol **2**, 423-9. (2001)

van't Wout, A.B., Kootstra, N.A., Mulder-Kampinga, G.A., Albrecht-van Lent, N., Scherpbier, H.J., Veenstra, J., Boer, K., Coutinho, R.A., Miedema, F.,Schuitemaker, H.: Macrophage-tropic variants initiate human immunodeficiency virus type 1 infection after sexual, parenteral, and vertical transmission. J Clin Invest **94**, 2060-7 (1994)

Veiga-Fernandes, H., Walter, U., Bourgeois, C., McLean, A.,Rocha, B.: Response of naive and memory CD8+ T cells to antigen stimulation in vivo. Nat Immunol **1**, 47-53 (2000)

Villarete, L., Somasundaram, T.,Ahmed, R.: Tissue-mediated selection of viral variants: correlation between glycoprotein mutation and growth in neuronal cells. J Virol **68**, 7490-6. (1994)

Wahl, L.M.,Nowak, M.A.: Adherence and drug resistance: predictions for therapy outcome. Proc R Soc Lond B Biol Sci **267**, 835-43 (2000)

Wei, X., Decker, J.M., Wang, S., Hui, H., Kappes, J.C., Wu, X., Salazar-Gonzalez, J.F., Salazar, M.G., Kilby, J.M., Saag, M.S., Komarova, N.L., Nowak, M.A., Hahn, B.H., Kwong, P.D.,Shaw, G.M.: Antibody neutralization and escape by HIV-1. Nature **422**, 307-12 (2003)

Wei, X.P., Ghosh, S.K., Taylor, M.E., Johnson, V.A., Emini, E.A., Deutsch, P., Lifson, J.D., Bonhoeffer, S., Nowak, M.A., Hahn, B.H., Saag, M.S.,Shaw, G.M.: Viral Dynamics in Human Immunodeficiency Virus Type-1 Infection. Nature **373**, 117-122 (1995)

Weiner, A.J., Geysen, H.M., Christopherson, C., Hall, J.E., Mason, T.J., Saracco, G., Bonino, F., Crawford, K., Marion, C.D., Crawford, K.A.,et al.: Evidence for immune selection of hepatitis C virus (HCV) putative envelope glycoprotein variants: potential role in chronic HCV infections. Proc Natl Acad Sci U S A **89**, 3468-72. (1992)

Welsh, R.M., Selin, L.K.,Razvi, E.S.: Role of apoptosis in the regulation of virus-induced T cell responses, immune suppression, and memory. J Cell Biochem **59**, 135-42 (1995)

Weninger, W., Manjunath, N.,von Andrian, U.H.: Migration and differentiation of CD8+ T cells. Immunol Rev **186**, 221-33. (2002)

Wherry, E.J., Blattman, J.N., Murali-Krishna, K., van der Most, R., Ahmed, R.: Viral persistence alters CD8 T-cell immunodominance and tissue distribution and results in distinct stages of functional impairment. J Virol **77**, 4911-27 (2003a)

Wherry, E.J., Teichgraber, V., Becker, T.C., Masopust, D., Kaech, S.M., Antia, R., von Andrian, U.H., Ahmed, R.: Lineage relationship and protective immunity of memory CD8 T cell subsets. Nat Immunol **4**, 225-34 (2003b)

Wodarz, D.: Cytotoxic T-lymphocyte memory, virus clearance and antigenic heterogeneity. Proc Biol Sci **268**, 429-36 (2001a)

Wodarz, D.: Mechanisms underlying antigen-specific CD8+ T cell homeostasis. Science **292**, 595. (2001b)

Wodarz, D.: Hepatitis C virus dynamics and pathology: the role of CTL and antibody responses. J Gen Virol **84**, 1743-50 (2003)

Wodarz, D., Arnaout, R.A., Nowak, M.A.,Lifson, J.D.: Transient antiretroviral treatment during acute SIV infection facilitates long-term control of the virus. Phil. Trans. Roy. Soc. Lond. B **355**, 1021-1029 (2000a)

Wodarz, D., Christensen, J.P.,Thomsen, A.R.: The importance of lytic and nonlytic immune responses in viral infections. Trends Immunol **23**, 194-200. (2002)

Wodarz, D., Hall, S.E., Usuku, K., Osame, M., Ogg, G.S., McMichael, A.J., Nowak, M.A.,Bangham, C.R.: Cytotoxic T-cell abundance and virus load in human immunodeficiency virus type 1 and human T-cell leukaemia virus type 1. Proc R Soc Lond B Biol Sci **268**, 1215-21. (2001)

Wodarz, D., Klenerman, P.,Nowak, M.A.: Dynamics of cytotoxic T-lymphocyte exhaustion. Proc R Soc Lond B Biol Sci **265**, 191-203 (1998)

Wodarz, D.,Krakauer, D.C.: Defining CTL-induced pathology: implications for HIV. Virology **274**, 94-104. (2000)

Wodarz, D., Lloyd, A.L., Jansen, V.A.A.,Nowak, M.A.: Dynamics of macrophage and T cell infection by HIV. Journal of Theoretical Biology **196**, 101-113 (1999)

Wodarz, D., May, R.M.,Nowak, M.A.: The role of antigen-independent persistence of memory CTL. International Immunology **12**, 467-477 (2000b)

Wodarz, D.,Nowak, M.A.: Specific therapy regimes could lead to long-term control of HIV. Proc. Natl. Acad. Sci. USA **96**, 14464-14469 (1999)

Wodarz, D.,Nowak, M.A.: CD8 memory, immunodominance and antigenic escape. Europ. J. Immunol. **30**, 2704-2712 (2000a)

Wodarz, D.,Nowak, M.A.: Immune responses and viral phenotype: do replication rate and cytopathogenicity influence virus load? Journal of Theoretical Medicine **2**, 113-127 (2000b)

Wodarz, D., Page, K.M., Arnaout, R.A., Thomsen, A.R., Lifson, J.D., Nowak, M.A.: A new theory of cytotoxic T-lymphocyte memory: implications for HIV treatment. Philos Trans R Soc Lond B Biol Sci **355**, 329-43 (2000c)

Wodarz, D.,Thomsen, A.R.: Does programmed CTL proliferation optimize virus control? Trends Immunol **26**, 305-10 (2005)

Wolinsky, S.M., Korber, B.T., Neumann, A.U., Daniels, M., Kunstman, K.J., Whetsell, A.J., Furtado, M.R., Cao, Y., Ho, D.D.,Safrit, J.T.: Adaptive evolution of human immunodeficiency virus-type 1 during the natural course of infection. Science **272**, 537-42 (1996)

Wong, J.K., Hezareh, M., Gunthard, H.F., Havlir, D.V., Ignacio, C.C., Spina, C.A.,Richman, D.D.: Recovery of replication-competent HIV despite prolonged suppression of plasma viremia. Science **278**, 1291-5 (1997)

Xiang, R., Lode, H.N., Gillies, S.D.,Reisfeld, R.A.: T cell memory against colon carcinoma is long-lived in the absence of antigen. J Immunol **163**, 3676-83 (1999)

Yewdell, J.W.,Bennink, J.R.: Immunodominance in major histocompatibility complex class I-restricted T lymphocyte responses. Annu Rev Immunol **17**, 51-88 (1999)

Zhang, L., Yu, W., He, T., Yu, J., Caffrey, R.E., Dalmasso, E.A., Fu, S., Pham, T., Mei, J., Ho, J.J., Zhang, W., Lopez, P.,Ho, D.D.: Contribution of human alpha-defensin 1, 2, and 3 to the anti-HIV-1 activity of CD8 antiviral factor. Science **298**, 995-1000 (2002)

Zimmerman, C., Brduscha-Riem, K., Blaser, C., Zinkernagel, R.M., Pircher, H.: Visualization, characterization, and turnover of CD8 memory T cells in virus-infected hosts. J Exp Med **183**, 1367-75 (1996)

Zinkernagel, R.: [Virus-induced immune deficiency disease AIDS: direct pathogenic effect of the virus or immunopathology?]. Verh Dtsch Ges Pathol **78**, 166-70 (1994)

Zinkernagel, R.M.: Immune protection vs. immunopathology vs. autoimmunity: a question of balance and of knowledge. Brain Pathol **3**, 115-21 (1993)

Zinkernagel, R.M.: Immunosuppression by a noncytolytic virus via T cell mediated immunopathology. Implication for AIDS. Adv Exp Med Biol **374**, 165-71 (1995)

Zinkernagel, R.M.: Immunology taught by viruses. Science **271**, 173-8 (1996)

Zinkernagel, R.M.: On immunological memory. Philos Trans R Soc Lond B Biol Sci **355**, 369-71. (2000a)

Zinkernagel, R.M.: What is missing in immunology to understand immunity? Nat Immunol **1**, 181-5. (2000b)

Zinkernagel, R.M.: On differences between immunity and immunological memory. Curr Opin Immunol **14**, 523-36. (2002a)

Zinkernagel, R.M.: Uncertainties - discrepancies in immunology. Immunol Rev **185**, 103-25. (2002b)

Zinkernagel, R.M., Althage, A.,Jensen, F.C.: Cell-mediated immune response to lymphocytic choriomeningitis and vaccinia virus in rats. J Immunol **119**, 1242-7. (1977)

Zinkernagel, R.M.,Hengartner, H.: T-cell-mediated immunopathology versus direct cytolysis by virus: implications for HIV and AIDS. Immunol Today **15**, 262-8 (1994)

Zinkernagel, R.M., Planz, O., Ehl, S., Battegay, M., Odermatt, B., Klenerman, P.,Hengartner, H.: General and specific immunosuppression caused by antiviral T-cell responses. Immunol Rev **168**, 305-15 (1999)

Index

adaptive immunity, 5, 6, 9, 41, 42, 55, 192
antibodies, 5, 6, 10, 16, 22, 41, 42, 48, 55, 114, 115, 123–136, 144, 148, 177, 184, 192
antigen presenting cell, 8, 9, 19, 20, 48, 56–59, 61–65, 70, 149, 156
APC, *see* antigen presenting cell

basic reproductive ratio of the virus, 27, 28
bistability, 152, 153, 155, 156, 162, 166, 170, 189

CD4 cell, *see* helper T cells
CD8 cell, *see* CTL
clearance of infection, 5, 9, 14, 22, 30, 35, 38, 39, 41, 42, 45, 47, 49, 50, 64–70, 85, 87, 88, 94, 99, 107, 116, 120, 128, 129, 133, 144, 147, 156
clonal expansion, 9, 10, 44, 55, 71, 74, 80
CMV, *see* cytomegalo virus
competition, 1, 71–73, 77, 80, 81, 84, 87, 93, 97, 125–128, 133, 134, 136, 140, 143, 162, 186–188, 193
  antibodies vs CTL, 126
contraction phase, 9, 43, 77, 82, 138
correlation between immunity and virus load, 163
CTL, 2, 5–18, 22, 24, 25, 28–53, 55–97, 99–111, 113–116, 118–145, 147–153, 155–170, 176–178, 181, 183, 192

aging, 96
competition, 71
exhaustion, 156
helper dependent vs independent, 158
homeostasis, 137
interactions with antibodies, 125
lytic vs nonlytic responses, 113
memory inflation, 76, 77, 81–84
programmed proliferation, 142, 143
CTL-induced pathology, 99
  in HIV infection, 108
  in LCMV infection, 103
  mathematical insights, 100
  role of lytic and nonlytic responses, 116
cytomegalo virus, 77, 79, 80
cytotoxic T lymphocyte, *see* CTL

dendritic cells, 8, 19, 20, 149

ecology, 1, 2, 138
effector molecules
  and CTL numbers, 138
  and immunodominance, 140
  role in VSV infection, 144
epitope, 5, 10, 72, 73, 75–77, 147
evolution, 2, 17, 20–24, 75, 120, 125, 127, 129–131, 133–136, 183, 184, 188, 189, 191–193
  immunological memory, 183
  of pathogen virulence, 189
experimental data/observations, 2, 13, 18, 25, 28, 38, 42, 43, 47, 51–53, 70, 84, 87, 93, 99, 100, 108, 111,

115, 118, 123, 124, 127, 129, 133, 137, 138, 142, 143, 145, 156, 174, 181

FAS, 9, 114, 123, 140

gamma herpes virus, 15, 47, 48, 66, 93, 94, 97

half-life of HIV infected cells, 18
HBV, see hepatitis B virus
HCV, see hepatitis C virus
helper T cells, 6, 8–11, 14–17, 19–24, 42, 47–49, 55–66, 68–70, 75, 103, 105–111, 122, 123, 135, 148–153, 156, 158–169, 174, 176, 178, 179, 181
  classical pathway, 61
  CD4-APC-CTL pathway, 57
  classical pathway, 57
  helper deficient hosts, 66
  impairment, 148
  importance of different pathways, 63
  mathematical model, 56
  mechanism of help, 56
hepatitis B virus, 10, 15, 100, 113, 124, 167, 179
hepatitis C virus, 10, 17, 53, 68, 96, 100, 126–129, 131, 133–136, 147, 149, 167–169, 178, 181
  antibody vs CTL in acute infection, 128
  antibody vs CTL in chronic infection, 129
  antigenic diversity, 133
  viral evolution and disease progression, 129, 131
HIV, see human immunodeficiency virus
HTLV-1, see human T cell leukemia virus
human immunodeficiency virus, 2, 4, 5, 10, 13–24, 53, 75, 96, 100, 108–111, 113, 124, 147–149, 158, 163, 165, 167–169, 174, 177, 178, 181
  antigenic escape, 21, 22
  biphasic virus decline on therapy, 19
  combination therapy, 17, 21
  compliance, 21
  diversity threshold, 22
  evolution and disease progression, 23
  evolution in vivo, 20
  HAART, 17
  HIV decay slopes, 18
  kinetics, 18, 19
  long-term nonprogressors, 16, 75, 158, 163, 167, 168
  mutation rate, 16
  natural history, 16
  protease inhibitors, 17
  resistance, 20, 21
  reverse transcriptase inhibitors, 17
  types of infected cells, 20
human T cell leukemia virus, 76, 120, 163

IFN, see interferon gamma
immune impairment, 147, 168
immune phenome, 96
immunodominance, 24, 71, 72, 74, 75, 77, 83, 84, 138, 140–143, 145
  and antigenic escape, 75
  in HIV infection, 75
  in murine CMV infection, 77
  mathematical model, 72
immunological memory, 9, 10, 14, 24, 30, 33–38, 41–53, 55–57, 63–66, 69, 72–77, 79–81, 83–97, 107, 137, 139, 143–145, 148–150, 157–162, 164, 168, 169, 176, 181, 183, 184, 186–193
  and pathogen competition, 188, 189
  and protection against reinfection, 49, 51
  and resolution of primary infection, 45, 47
  duration, 188, 191
  evolution, 183
  role of antigen, 50, 51
immunopathology, 99, 115, 120, 131
influenza virus, 15, 47, 49, 93, 94, 97, 123, 124, 187
innate immunity, 77, 79
integrase, 4, 5
interferon gamma, 9, 103–108, 113, 114, 120, 121, 137–145, 156, 168

knockout mice, 14, 47, 114, 115, 138–140, 142–145

latent infection, 12, 18, 47, 77, 93, 147
LCMV, see lymphocytic choriomeningitis virus
lymphocytic choriomeningitis virus, 13–15, 47, 48, 52, 53, 66, 69, 70, 99–101, 103–106, 108, 110, 111, 113, 115, 119–121, 124, 137, 140, 149, 156–158, 166
lytic immunity, 113, 115, 116, 125, 127, 136

macrophages, 8, 19, 20, 124, 145, 156
major histocompatibility complex, 7–9, 47, 60, 73, 121, 140, 142
mathematical models
  antibody-CTL interactions, 126
  basic CTL models, 29, 31, 32
  basic immune impairment model, 149
  basic virus dynamics, 26, 28
  CTL-antibody competition, 126
  epidemiological, 184, 186
  general class of impairment models, 153
  helper cells, 56
  helper dependent vs independent, 158
  immunodominance, 72, 78
  memory CTL models, 33
  multiple infections, 86
  programmed CTL proliferation models, 35
  robustness, 40, 132, 144, 155, 166, 169
memory, see immunological memory
MHC, see major histocompatibility complex
multiple infections, 85
  and antibiotic treatment, 94
  and memory collapse, 90
  and the immune phenome, 96
  and vaccination, 95
  and virus control, 87
  gamma herpes and influenza viruses, 93

natural killer cells, 77–84
nonlinear, 18

nonlytic immunity, 113, 115, 116, 119, 125, 127, 136

ordinary differential equations, 26, 29, 31, 33, 35, 64, 72, 78, 86, 109, 116, 126, 150, 153, 159, 163, 185, 186

perforin, 9, 14, 114, 115, 120, 123, 137–145
persistent infection, 10, 11, 17, 24, 25, 34, 36, 38–40, 43, 45, 46, 64, 66, 70, 80, 88, 90, 96, 97, 101, 127–131, 133, 134, 137, 138, 144, 149, 156, 158, 168, 169
predator, 1, 29, 79, 87, 138, 142, 144, 145, 184
primary infection, 41, 45–47, 50, 51, 53, 56, 66, 77, 87, 107, 174, 176, 191
programmed CTL proliferation, 35–40, 67, 79, 142–144
protease, 4, 5, 17
protease inhibitors, 5, 17

quasiequilibrium, 34, 38, 46, 73, 75

R_0, see basic reproductive ratio of the virus
rechallenge, see secondary infection
retrovirus, 5
reverse transcriptase, 4, 5, 16, 17, 20
reverse transcriptase inhibitors, 5, 17

secondary infection, 9, 41, 49, 51, 52, 64, 66, 85, 97, 157, 174–177
simian immunodeficiency virus, 22, 149, 164, 165, 168, 174–177, 181
SIV, see simian immunodeficiency virus
specificity, 8, 21, 22

T cell, see CTL; helper T cells
T cell receptor, 6, 7, 10, 55, 60
therapy
  boosting immunity, 167
  boosting immunity in HCV, 178
  boosting immunity in SIV/HIV, 174
  protease inhibitors, 5, 17
  protection against reinfection, 177
  protection against SIV reinfection, 175
  reverse transcriptase inhibitors, 5, 17

structured treatment interruptions, 179
T cell dynamics, 170

vesicular stomatitis virus, 15, 52, 121, 123, 124, 144, 145
virus
  cytopathicity/cytotoxicity, 27, 121–124

life cycle, 3
replication rate, 4, 5, 9, 12, 14, 17–19, 21, 23, 25, 26, 31, 32, 51, 52, 64, 67–70, 73, 76–78, 87, 93, 100–111, 113–124, 131, 134, 137, 140–142, 151–157, 159–162, 166, 169–171, 173, 174, 176, 177

VSV, *see* vesicular stomatitis virus

Interdisciplinary Applied Mathematics

1. *Gutzwiller:* Chaos in Classical and Quantum Mechanics
2. *Wiggins:* Chaotic Transport in Dynamical Systems
3. *Joseph/Renardy:* Fundamentals of Two-Fluid Dynamics: Part I: Mathematical Theory and Applications
4. *Joseph/Renardy:* Fundamentals of Two-Fluid Dynamics: Part II: Lubricated Transport, Drops and Miscible Liquids
5. *Seydel:* Practical Bifurcation and Stability Analysis: From Equilibrium to Chaos
6. *Hornung:* Homogenization and Porous Media
7. *Simo/Hughes:* Computational Inelasticity
8. *Keener/Sneyd:* Mathematical Physiology
9. *Han/Reddy:* Plasticity: Mathematical Theory and Numerical Analysis
10. *Sastry:* Nonlinear Systems: Analysis, Stability, and Control
11. *McCarthy:* Geometric Design of Linkages
12. *Winfree:* The Geometry of Biological Time (Second Edition)
13. *Bleistein/Cohen/Stockwell:* Mathematics of Multidimensional Seismic Imaging, Migration, and Inversion
14. *Okubo/Levin:* Diffusion and Ecological Problems: Modern Perspectives (Second Edition)
15. *Logan:* Transport Modeling in Hydrogeochemical Systems
16. *Torquato:* Random Heterogeneous Materials: Microstructure and Macroscopic Properties
17. *Murray:* Mathematical Biology I: An Introduction (Third Edition)
18. *Murray:* Mathematical Biology II: Spatial Models and Biomedical Applications (Third Edition)
19. *Kimmel/Axelrod:* Branching Processes in Biology
20. *Fall/Marland/Wagner/Tyson (Editors):* Computational Cell Biology
21. *Schlick:* Molecular Modeling and Simulation: An Interdisciplinary Guide
22. *Sahimi:* Heterogeneous Materials I: Linear Transport and Optical Properties
23. *Sahimi:* Heterogeneous Materials II: Nonlinear and Breakdown Properties and Atomistic Modeling
24. *Bloch:* Nonholonomic Mechanics and Control
25. *Beuter/Glass/Mackey/Titcombe (Editors):* Nonlinear Dynamics in Physiology and Medicine
26. *Ma/Soatto/Košecká/Sastry:* An Invitation to 3-D Vision: From Images to Geometric Models
27. *Ewens:* Mathematical Population Genetics (Second Edition)
28. *Wyatt:* Quantum Dynamics with Trajectories: Introduction to Quantum Hydrodynamics
29. *Karniadakis/Beskok/Aluru:* Microflows and Nanoflows: Fundamentals and Simulation
30. *Macheras:* Modeling in Biopharmaceutics, Pharmacokinetics and Pharmacodynamics
31. *Samelson/Wiggins:* Lagrangian Transport in Geophysical Jets and Waves
32. *Wodarz:* Killer Cell Dynamics: Mathematical and Computational Approaches to Immunology

Printed in the United States of America